Front Cover Photograph
Courtesy of the U.S. Marine Corps Recruiting Command

Rear Cover Photograph
Courtesy of Leeson Photography (and enhanced by the U.S. Marine Corps Recruiting Command)

– Books by Marion Sturkey –

BONNIE-SUE: A Marine Corps Helicopter Squadron in Vietnam
ISBN: 0-9650814-2-7 (509 pages)

Warrior Culture of the U.S. Marines (first edition)
ISBN: 0-9650814-5-1 (207 pages)

Warrior Culture of the U.S. Marines (second edition)
ISBN: 0-9650814-1-9 (212 pages)

Murphy's Laws of Combat (first edition)
ISBN: 0-9650814-4-3 (241 pages)

Murphy's Laws of Combat (second edition)
ISBN: 978-0-9650814-6-7 (366 pages)

MAYDAY: Accident Reports and Voice Transcripts from Airline Crash Investigations
ISBN: 978-0-9650814-3-6 (461 pages)

MID-AIR: Accident Reports and Voice Transcripts from Military and Airline Mid-Air Collisions
ISBN: 978-0-9650814-7-4 (477 pages)

GONE, BUT NOT FORGOTTEN: An Introduction
LCCCN 88-92649 regional historical interest only (42 pages)

GONE, BUT NOT FORGOTTEN
LCCCN 88-92560 regional historical interest only (679 pages)

(coming soon) *Military Monuments in South Carolina*
ISBN: 978-0-9650814-8-1

Murphy's Laws of Combat

The American Warrior's Guide to ***Staying Alive*** in Battle

Marion F. Sturkey

Heritage Press International

Murphy's Laws of Combat

Library of Congress Control Number: 2009921056

ISBN: 978-0-9650814-6-7

Second Edition

Heritage Press International
204 Jefferson Street
P.O. Box 333
Plum Branch, SC 29845 USA

USMCpress.com
MarionS@wctel.net

Manufactured in the United States of America

for
American Patriots
who have
served in the
U.S. Armed Forces

Table of Contents

– PART ONE –
– Murphy's Basic Laws for All Mankind –
(Civil Satire)

– PART TWO –
– Murphy's Laws of Combat –
(Military Satire)

(continued on next page)

(Table of Contents, continued)

(continued on next page)

(Table of Contents, continued)

– PART THREE –
– Heritage of the American Warrior –
(Tribute to U.S. Warriors)

Acknowledgments

Grateful acknowledgment is extended to the late Capt. Edward A. Murphy Jr. (1917-1987), USAF, the "father" of ***Murphy's Law***. In 1949, Capt. Murphy first uttered the prophetic statement which became the basis for this perpetual thorn in the foot of humanity.

Grateful acknowledgment is extended to the U.S. Marine Corps Recruiting Command for assistance with regard to the Recruiting Poster images depicted on the front and rear covers of this book.

Grateful acknowledgment is extended to the owners of Leeson Photography for assistance with regard to the photograph of a wolf (later enhanced by the U.S. Marine Corps Recruiting Command) depicted on the rear cover of this book.

Grateful acknowledgment is extended to each American Patriot who has served in the Armed Forces of the United States of America.

Grateful acknowledgment is extended to the following governmental entities for making their photographs and graphic images available for use in this project:

- U.S. National Defense University
- U.S. National War College
- U.S. Department of Defense
- U.S. Army
- U.S. Air Force
- U.S. Marine Corps
- U.S. Navy
- U.S. Library of Congress
- U.S. National Aeronautics and Space Administration (NASA)
- U.S. National Archives and Records Administration

Heads-Up for Readers

Military Abbreviations: The individual branches of Armed Forces most frequently referenced in this book are identified by abbreviations as follows:

USA .... United States Army
USN .... United States Navy
USMC ..... United States Marine Corps
USAF ...... United States Air Force
CSA .... Confederate States Army
RN .... Royal Navy (British)
RAF .... Royal Air Force (British)

Civilian Sources: The title, position, occupation, or claim-to-fame of civilian sources is stated when known, for example: (1) U.S. Senator, (2) French military strategist, (3) Italian philosopher, etc.

On Active Duty? With respect to rank and military affiliation, this book makes no distinction between military warriors on active duty, military retirees, military veterans, military reservists, etc.

Intentional Duplications: Several quotations from various other chapters in this book are intentionally duplicated in the chapter, "Murphy's Military Superlatives."

Warning!

Politically Impossible!

The content of ***PART TWO*** in this book is tailored for professional red-blooded American Warriors. Readers will find no politically correct psycho-babble.

Any brain-dead degenerate liberal who is offended by this warrior ethos should look for something else to read, such as *Floral Arrangements for All Occasions*.

Those who dance are
thought to be quite insane
by those who can not
hear the music.

– PART ONE –

Murphy's Basic Laws for All Mankind

(Civil Satire)

The Origin of Murphy's Law

Murphy's Law is not some figment of creative imagination, for this timeless truth has existed since time immemorial. Yet, it was not officially named and published until the Twentieth Century.

After World War II the United States aerospace experiments had shifted into high gear. U.S. Air Force Project MX-981 used a rocket-sled test program. On the sled, named the *Sonic Wind*, volunteers explored the G-force limits a human body can withstand.

Murphy's Law is not some figment of creative imagination.

Capt. Edward A. Murphy Jr. (1917-1987), a brilliant Air Force aerospace engineer, worked on these experiments at what is now Edwards Air Force Base in California. One sunny morning in 1949, Murphy's staff strapped a volunteer, Maj. John P. Stapp, USAF, into the sled. A technician attached 16 accelerometers to the sled. Murphy had estimated that Stapp would be subjected to 40 G's for 1.7 seconds.

Volunteers explored the G-force limits a human body can withstand.

When all was ready the rockets ignited with a deafening roar. The sled blasted down the rails, propelled by a white-hot furnace of flame from the rocket. Then the rocket motor switched off, and the speeding *Sonic Wind* screeched to a rapid stop.

Medical aides ran to help Maj. Stapp, who was semiconscious and bleeding from several body orifices. The aides spirited him away for a medical checkup. Murphy's technicians then checked the sled, and to their surprise they found that all 16 accelerometers registered *zero*. What had gone wrong?

The technician who installed the accelerometers had carefully installed each one – ***backwards!***

Murphy quickly found the problem. The technician who installed the accelerometers had carefully installed each one – ***backwards!*** Murphy reacted as though struck by lightning. Leaping to his feet, he looked up toward Heaven and lamented:

> If there are two ways to do something, and one of those ways will cause a catastrophe, some idiot will do it that way!

George Nichols, the project manager, quickly scribbled Murphy's lamentation into his notebook. As an afterthought he jotted-down an off-the-cuff caption: "Murphy's Law."

The next day at a press conference a revived Maj. Stapp talked with the news media. He joked that despite "Murphy's Law" the rocket-sled project had a good safety record.

George Nichols and Maj. Stapp had planted the seed of truth. Within a few weeks ***Murphy's Law*** began appearing in aerospace advertisements. The world-renowned Flight Safety Foundation soon published an abbreviated version:

> – Murphy's Law –
>
> If something can go wrong, it *will* go wrong!

Within months Murphy's Law migrated to various technology ventures and engineering firms. Soon it found its way onto college campuses and into the business world. It grew and grew, mutating and evolving all the while.

In 1958, Webster's Dictionary included Murphy's Law, bowing to the inherent truth of this perpetual thorn in the foot of humanity. Murphy's Law eventually spanned oceans and cultures, achieving immortality in the popular imagination of all mankind.

Murphy's Basic Laws for All Mankind

Murphy's basic laws govern all human endeavors, military and civilian alike. Historically, these fundamental truths formed a ball and chain around the foot of progress. Tragically these laws are constant. There is no appeal, and there is no escape. Therefore, before a warrior delves into laws and truisms that govern military arts, he first must gain an understanding of bedrock principles that plague all mankind.

> These fundamental truths formed a ball and chain around the foot of progress.

Although Edward Murphy died in 1987, his philosophy will live forever. Murphy's Law will never die. So, in a sense, one can say that ***Murphy is alive and well***. In that context he sets forth his fundamental principles for humanity in three simple categories:

1. Murphy's General Laws
2. Murphy's Technology Laws
3. Murphy's Technological Advice

– Murphy's General Laws –

If something *can* go wrong, it *will* go wrong!

> You *never* run out of things that can go wrong.

If something simply can't go wrong, it will go wrong anyway.

Things that go wrong always do so when you least expect it.

When two or more things go wrong, they will do so all at once.

You *never* run out of things that can go wrong.

Of two possible events, only the undesired event will occur.

Any event which would create chaos if it did occur, will occur.

Left to themselves, things go from bad to worse.

Nothing is ever so bad that it can't get worse.

Left to themselves, things go from bad to worse.

Sooner or later, the worst possible condition will occur.

If there is a worst time for something to go wrong, it always will go wrong then.

If you determine there are four possible ways in which something can go wrong, and circumvent those four ways, then a fifth way, unprepared for, will develop.

Anything that begins well, ends badly.

Things always get worse before they get better – and then they *don't* get better.

Things aren't usually quite as bad as they seem – they're worse.

Things always get worse under pressure.

Things always get *much* worse at night.

After going from bad to worse, the cycle will repeat itself.

The only genuine change is for the worse.

Progress is when things get worse at a slower rate.

Anything that begins well, ends badly.

Anything that begins badly, ends worse.

There is no limit to how bad things can get.

Two wrongs are only the beginning.

Nature always sides with the hidden flaw.

Nothing is as easy as it looks.

Anything *may* be possible, but nothing is easy.

Everything takes longer than you anticipate.

Nothing can be done in the allotted time.

Anything is easier to *get into* than to *get out of.*

Nothing is impossible for those who want somebody else to do it.

You can't ever simply do *one* thing – whenever you set out to do something, *something else* always must be done first.

Nature always sides with the hidden flaw.

Unfortunately, the hidden flaw never remains hidden for long.

Anything *may* be possible, but nothing is easy.

Yet, if there is a hidden *solution*, it will remain hidden.

Nothing is so simple that it can't be misunderstood.

Nothing is so simple that it can't be misunderstood.

Each failure further reinforces the failure mind-set.

The degree of failure will be *directly* proportional to the need for success.

Nothing succeeds like failure, and when failure rains, it pours.

The most crucial problems have no genuine solutions.

If you think the problem is bad now, just wait until it's solved!

Nothing is devoid of side effects, nothing ever goes away, and there is no free lunch.

In the unlikely event there is a correct solution to a problem, it will magnify the problem.

Each problem solved breeds a new and more complex problem.

All complex problems have easy-to-understand wrong answers.

Every person has a unique sure-fire solution that won't work.

The correct answer to a complex problem will be (1) partly right, or (2) partly wrong, but (3) never simply right or wrong.

The number of hypotheses that can explain any given phenomenon is always infinite.

Nothing is devoid of side effects, nothing ever goes away, and there is no free lunch.

The bird in hand is usually dead.

Most things, if left to chance, would turn out much better.

Nothing is as good as it seemed beforehand.

Things that are supposed to happen won't happen – but the most unexpected thing *will* happen.

Most things, if left to chance, would turn out much better.

Everything *depends*, and everything *breaks down*.

Sooner or later, things always go wrong. This big CH-46 helicopter has plopped down into a river (photo courtesy of U.S. Marine Corps).

Nothing is ever done for the right reason.

There are no absolute truths, and 90 percent of everything is crud.

There's always a correct solution – but it never works.

There's always a correct solution – but it never works.

Most of what you think is right is wrong.

Plus, there's always something more that's wrong.

Those who aren't confused aren't well informed.

If opportunity knocks, it will be at the least opportune moment.

You get the most of what you need the least.

Those who aren't confused aren't well informed.

Presumed variables won't vary, and change is the only constant.

The cream rises, and rises, and rises – until it sours.

Those who fail to study history will repeat its errors. Those who do study history will find other ways to err.

– Murphy's Technology Laws –

Reality is an illusion.

Most advanced technology is indistinguishable from magic.

Any technology that depends upon human reliability is unreliable.

Any object in motion is moving in the wrong direction.

Any object at rest is resting in the wrong place.

Anticipated events never occur, and if they seem to occur, they never live up to expectations.

Any object in motion is moving in the wrong direction.

Any unexpected sequence of *events* will be followed by an unexpected sequence of *trouble*.

If there's a fifty-fifty chance a process will go wrong, it will go wrong nine times out of ten.

Any object at rest is resting in the wrong place.

Once a job is irrevocably screwed up, anything done to try to improve it will only make it worse.

The most technically advanced and expensive machines will become piles of burned rubble (photo by the author, Marion Sturkey).

The reliability of any system is *inversely* proportional to the square of its complexity.

If the assumptions are wrong, so are the conclusions.

If a scientific process absolutely can not go wrong, it will anyway.

In any crisis, if the experts all agree, they're all wrong.

In any crisis, whatever hits the fan is never evenly distributed.

A *good scapegoat* is much better than a *good solution.*

An ounce of applicability is worth a ton of abstract theory.

Sadly, an ounce of image is valued more than a ton of competence.

Nothing is ever right in nature. Therefore, if something looks natural, it's always wrong.

There is always a solution – simple, plausible, and wrong.

A *good scapegoat* is much better than a *good solution.*

If a complex experiment seemed to work, something went wrong.

If the experiment *really* worked, it was the *wrong* experiment.

Indecision is the key to flexibility.

Logic is a method for reaching the wrong conclusion with confidence.

Most of the time, it's better to be *lucky* than *logical.*

Most of the time, it's better to be *lucky* than *logical.*

Nonetheless, no logical act goes unpunished.

When things go wrong, the urgency of repair work will be *inversely* proportional to the availability of spare parts.

If it *can* be borrowed and broken, it *will* be borrowed and broken.

If it *can* be borrowed and broken, it *will* be borrowed and broken.

If something can break, it will – but only after the warranty expires.

If things are going exactly as planned, you've overlooked something.

If things *seem* to be going well – wait five minutes.

If there's a possibility of several things going wrong, and only one actually goes wrong, it will be the one that will (1) cause the most damage and (2) cost the most money.

Everything takes (1) more time, and (2) more money.

Indecision is the key to flexibility.

If it weren't for the last minute, nothing ever would get done.

In any process, the probability of something happening is *inversely* proportional to its desirability.

Undetectable errors are infinite in quantity and variety.

Nobody perceives anything with total accuracy.

When any device is used to its full potential, it will break.

When any device is used to its full potential, it will break.

Each falling object will fall in the precise manner necessary to do the most damage.

If only one object falls, it will be the most delicate component.

Sooner or later, all delicate components will be dropped.

Any tool, when dropped, will roll into the most inaccessible corner.

If a scientific project ever begins to go quite well, some idiot will begin to experiment detrimentally.

Any system, however complicated, if examined in precisely the right way, will become even more complicated.

In scientific circles, all *reasoning* consists of a desperate search for data to justify believing what they already believe.

Try as you may, you can't tell which way the train went by looking at the railroad tracks.

Things put together will fall apart sooner or later – usually sooner.

Creativity varies *inversely* with the number of creative participants.

Any system, however complicated, if examined in precisely the right way, will become even more complicated.

If you tell a man there are 300 billion stars in the universe, he will believe you. If you tell him the bench has wet paint on it, he will touch it to make sure.

Man will occasionally stumble over the truth. Unfortunately he will usually pick himself up, dust himself off, and continue on as though nothing has happened.

– Murphy's Technological Advice –

Don't merely *believe* in miracles, *rely* on them.

If at first you don't succeed, you won't *ever* succeed. Destroy all evidence that you tried.

If at first you don't succeed, the only way to avoid further failure is to quit trying.

If at first you don't succeed, the only way to avoid further failure is to quit trying.

All truly great discoveries are made by mistake.

Any great discovery is more likely to be exploited by the wicked than applied by the virtuous.

Never let facts disrupt a carefully thought-out bad decision.

After all logical efforts have failed, read the instructions.

If a delicate machine jams, don't force it – just try a bigger hammer.

Most electrical devices work better after they're plugged-in.

If a delicate machine jams, don't force it – just try a bigger hammer.

The first rule of intelligent tinkering is to *save all parts*.

If you tinker with something long enough, it will break.

To pick the smartest person, pick the one who predicts the project will (1) take too long, (2) cost too much, and (3) fail.

When working on the solution to a complex problem, it's useful to already know the answer.

When you don't know what you're doing, do it neatly.

No experiment is a total failure – it can serve as a bad example.

When working on the solution to a complex problem, it's useful to already know the answer.

If a wrong thing is done long enough, it becomes right.

Any *theory* will fit any *facts* – if you make enough assumptions.

If it can't be expressed numerically, it's a theory, not a fact.

Enough research will support the most ludicrous theory.

Any *theory* will fit any *facts* – if you make enough assumptions.

Attempt the impossible, and you'll *prove* it's impossible.

The sooner you fall behind, the more time you'll have to catch up.

If you think you see the light at the end of the tunnel, it usually will be the headlight of an oncoming train.

Attempt the impossible, and you'll *prove* it's impossible.

Letting the cat out of the bag is much easier than putting the cat back in.

If you open a can of worms, the only way to re-can them is with a bigger and more expensive can.

Lost things will be found in the last place you look (*think* about it).

The laws of physics prove that matter can be neither created nor destroyed. But, it can be *lost*.

Luck is an acceptable substitute for competence, but only for those who can be *consistently* lucky.

The laws of physics prove that matter can be neither created nor destroyed. But, it can be *lost*.

If the shoe fits, you're not allowing for growth.

If you get the wrong answer, try multiplying by the page number.

When all else has failed, revel in the absurdity of it all!

In life, no matter how hard you try, (1) you can't win, (2) you can't break even, and (3) you can't quit.

After all is said and done, a lot more has been said than done.

When all else has failed, revel in the absurdity of it all!

– PART TWO –

Murphy's Laws of Combat

(Military Satire)

Murphy's Laws of Combat for Infantry

These laws of combat apply to all warriors. Yet, they are of prime concern to the population control specialists, the ***Magnificent Grunts***. These time-tested laws of combat never change, so warriors of today can learn from fatal mistakes made by warriors of the past. Otherwise, a modern warrior won't survive long enough to learn to survive permanently.

A sharpened rock, strapped onto a club, became a sophisticated war-ax. Hi-tech!

Although laws of combat never change, weapons do. In the beginning, combatants used their fists and teeth. Later they graduated to clubs and big rocks. Pretty soon a sharpened stick evolved into a spear. Then a sharpened rock, strapped onto a club, became a sophisticated war-ax. Hi-tech! Swords and shields followed. The longbow and arrows came next, and later the crossbow became an even more lethal and accurate killing machine. Then the Chinese stumbled across gunpowder, and the rest is history.

At his beck and call are flying machines of incredible speed that can rain aerial death and destruction down upon the bad guys.

The modern warrior charges into battle armed with a dizzying array of guns, mines, rockets, missiles, and electronic smart weapons. He can race overland from one battlefield to the next, protected inside his lethal armored chariot. At his beck and call are flying machines of incredible speed that can rain aerial death and destruction down upon the bad guys.

The modern warrior charges into battle armed with a dizzying array of guns, mines, rockets, missiles, and electronic smart weapons.

Warriors, handed down

through the ages, here are ***Murphy's Laws of Combat for Infantry***. These are your axioms and principles for staying alive in battle, so ignore them at your own peril. Murphy presents these priceless pearls of military wisdom in five categories:

1. Murphy's Infantry Philosophy
2. Murphy's Infantry Tactics
3. Murphy's Infantry Precautions
4. Murphy's Infantry Ironies
5. Murphy's Barracks Wisdom

– Murphy's Infantry Philosophy –

If you're allergic to lead, you'd be wise to avoid combat.

In combat, any warrior who doesn't consider himself the best in the game is in the *wrong* game.

War doesn't decide who's *right*. War decides who's *left*.

War doesn't decide who's *right*. War decides who's *left*.

There's one basic rule in warfare – the winner gets to make up the rules.

Remember that *diplomacy* is the art of saying "nice doggie" until you can find a bigger rock.

If you can win without fighting, that's *preferable*. But, it's also *harder*, and the bad guys may not cooperate.

Sooner or later in life, every warrior has to die. The nifty trick is to *die young* – as late as possible.

Warriors live a rough life. Only the young die good.

The only warriors fit to live are those who aren't afraid to die.

Fools trust their enemies. Skepticism is the mother of survival.

Medals are OK. Having all your warrior brothers alive is better.

For civilians, to *err* is human, and to *forgive* is divine. For warriors in combat, neither is acceptable.

In combat, no matter how bad it gets, having all your body parts intact and functioning makes it a good day.

If you win, nothing hurts, and you're entitled to the spoils of war. If you lose, you won't care.

Don't believe what they taught you in grammar school. In the real world and in combat, *violence* solves *everything*.

Don't believe what they taught you in grammar school. In the real world and in combat, *violence* solves *everything*.

To triumph in war, like in love, you must initiate contact.

It's usually easier to forgive an enemy *after* you've killed him.

Only imbeciles and fools fight fair – and won't do so for long.

Fools trust their enemies. Skepticism is the mother of survival.

The "Law of the Bayonet" says that warriors who fight with *automatic weapons* will win.

Those who "live by the sword" will die when they face those who fight with automatic weapons.

Even the Boy Scouts have figured it out: "Be Prepared."

– Murphy's Eternal Truths for Grunts –

The raw intensity of a War Story will be ***inversely*** proportional to the combat experience of the storyteller.

The probability of diarrhea will be ***directly*** proportional to the square of the thistle content of the local vegetation.

The urgency of the need to urinate will be (1) ***inversely*** proportional to the temperature, and (2) ***directly*** proportional to the layers of clothing you'll have to remove.

The number of mosquitoes present will be ***inversely*** proportional to your remaining amount of repellant.

The weight of your pack will be ***directly*** proportional to the cube of the time you've been humping it.

The severity of inclement weather will be ***directly*** proportional to the amount of time you'll be out in it.

When geared-up for combat, the severity of your itch will be ***inversely*** proportional to the length of your reach.

The complexity of your electronic equipment will be ***inversely*** proportional to the IQ of your civilian instructors.

During a firefight, the seriousness of your wound will be ***directly*** proportional to the distance to the nearest deep hole.

During a firefight, the difficulty of hearing a shouted order will be ***directly*** proportional to the consequences for failing to carry it out.

During a firefight, the intensity of enemy fire will be ***directly*** proportional to the curiousness of the enemy's target.

Forget that "the bigger they are, the harder they fall" foolishness. The bigger the bad guys are, the harder they punch, choke, and kick.

When you've run out of everything except the bad guys, you're definitely in combat.

When a warrior buddies-up for combat, he should avoid (1) pacifists and (2) cowards – those who think with their legs.

Warriors should never trust the sniveling media whores. You have to "read between the lies."

Warriors should never take part in a political war, where adversaries only "shoot from the lip."

{Know warfare, ***know*** peace and safety;
{No knowledge of warfare, ***no*** peace and safety.

Hot garrison chow is best. Hot field rations are better than cold field rations. Cold field rations are better than no food at all. Yet, no food at all is better than a cold rice ball a day, even though it may have little pieces of fish-head in it. ***Never surrender!***

A warrior's pack, however heavy, is lighter than a POW's chains.

A warrior's pack, however heavy, is lighter than a POW's chains.

Between firefights, *care packages* from home are great. Always share everything, even the pound cake and cookies.

Without daily resupply, neither colonels nor corporals are worth much.

In combat, air superiority isn't a luxury. Nonetheless, no one has yet invented any type of *flying* that will prevent a Grunt from *walking*.

If you're convinced that you'll lose, you're probably right.

If you're convinced that you'll lose, you're probably right.

In combat, last guys don't finish nice.

In combat, last guys don't finish nice.

Day or night, incoming fire always has right-of-way.

Two warriors charge the bad guys (photo courtesy of U.S. Marine Corps).

Those who beat their swords into plowshares will end up plowing for those who didn't.

Artillery lends dignity to what otherwise would be a vulgar brawl.

C-4 can convert a dull day into lots of fun.

Warriors in combat in the tropics never wear underwear, and only they can understand why.

Those who beat their swords into plowshares will end up plowing for those who didn't.

If they're shooting at you, it's a *high intensity conflict*.

Always make sure somebody has a P-38. For you new guys, a P-38 is a can-opener – and a lot more!

In combat, winning isn't always all it's cracked up to be. The only genuine winner in the War of 1812 was Tchaikovsky – who wouldn't be born for another 28 years.

In time of war, Hell hath no fury like a cut-and-run pacifist.

In combat, bad news often arrives in human waves.

Warriors know that letters from home aren't always so great. Both living and dying often hurt a lot.

In time of war, Hell hath no fury like a cut-and-run pacifist.

If the sheltered REMFs are happy, the warriors in combat likely don't have what they need.

The farther the REMF storyteller was from the battle, the thicker the flak will be in his story.

In combat, discretion is the *bitter* part of valor.

Unfortunately, combat isn't an efficient teacher. It gives the final test before presenting the lesson.

In combat a hero is no more brave than anyone else. He's merely brave a few minutes longer.

He who dies with the most toys is, nonetheless, dead.

When warriors take a calculated risk in a firefight, there usually are very few calculations.

In combat the only perfect science is called *hindsight*.

He who dies with the most toys is, nonetheless, dead.

In combat the only perfect science is called *hindsight*.

It's true that "experts" and "professionals" have their place, but don't get married to their ideas. Remember that (1) professionals built the *Titanic*, while (2) rank amateurs built the ark.

After the battle, life is valuable, but honor is priceless.

Remember, if all our warrior brothers don't come home, none of us can fully come home.

– Murphy's Infantry Tactics –

Do unto the enemy, and do it *first*.

If you don't strike first, you'll be the first struck.

Speak softly, but *forget* the big stick. Carry an automatic weapon.

In combat, true happiness is *always* an automatic weapon.

Always kill as many bad guys as you can. The ones you miss today may not miss you tomorrow.

If you can avoid it, never get into a fight without at least four times as much ammo as the bad guys.

It's better for you to hump extra ammo than for your buddy to fill out your Graves Registration paperwork.

Ammo is relatively cheap. Your life is not. In combat you can *never* hump too much ammo.

In combat, true happiness is *always* an automatic weapon.

If you have extra ammo, even in a firefight, share quickly. You may be on the short end the next time around.

If you can avoid it, never get into a fair fight.

– Murphy's 12 Rules for Firefights –

1. Never be the idiot who shows up armed only with a knife.
2. Bring an ***automatic weapon***. Better yet, bring *two*.
3. Bring all your friends, with all *their* automatic weapons.
4. Bring ***lots and lots of ammo*** – it's cheap life insurance.
5. If one of your weapons is a handgun, make sure its caliber begins with the numeral "4" or greater.
6. Make sure your weapons will fire ***every time***. If angel-pee causes your weapons to jam, you'll be terminally SOL.
7. Have a good plan. Have a good back-up plan.
8. Smoke and loud noise don't kill. ***Only hits count***.
9. The faster you shoot the bad guys, the less shot you'll get.
10. Any bad guy worth shooting is worth shooting several times. Play it safe and ***double-tap*** each bad guy.
11. The "hey-diddle-diddle" tactic works only in the movies, so be sneaky, always cheat, always win – and stay alive.
12. ***Taking prisoners*** is (1) a waste of your time, (2) troublesome, (3) counterproductive, and (4) ***not recommended***. Always double-tap all the bad guys.

In any firefight, always pace yourself. Otherwise, sooner or later you *will* run out of ammo – usually at the worst possible time.

In combat, never look back unless you intend to go that way.

The more ammo you have, the better. Nonetheless, it may spoil your day if, in a firefight, the ammo you have the *most of* is for the type of weapon you have the *least of*.

The "King of Battle" in action (photo courtesy of U.S. Marine Corps).

When given a choice, fight smarter, not harder.

Do what the enemy doesn't want you to do.

When you have the enemy on the ground, kick him.

If at first your well-planned attack doesn't succeed, don't try *again*. Try something *different*.

If trying something different doesn't work, radio for an airstrike.

In a firefight, nobody cares what you did yesterday, or what you'll do tomorrow. The thing that matters is what you're doing ***right now!***

In combat, never look back unless you intend to go that way.

In combat, the farther you are from your warrior brothers-in-arms, the less able they are to help you when you need them the most.

In a firefight, if you see two colonels conferring, you likely have fallen back a little too far.

If in command in a crisis, give all orders verbally. Never create a document that could wind up in a "Pearl Harbor" file.

If attacked by a fanatical and numerically superior enemy, it may be helpful to ponder: "How would the Lone Ranger handle this?"

Curious-looking objects attract fire. Never lurk behind one.

Cover your fellow warriors, so they'll be around to cover you.

On patrol and ambush, (1) never stand when you can sit, (2) never sit when you can lie down, (3) never stay awake when you can sleep, and (4) have a good bowel movement whenever you can.

In combat, hang on to your gear. If you drop it during a firefight, you often can find your canteen and E-tool right at your feet, but your ammo and grenades are probably lost forever.

Curious-looking objects attract fire. Never lurk behind one.

If you lose contact with the enemy, remember to look behind you.

When outnumbered in a firefight, shooting the bad guys is far more important than radioing your plight to some REMF, miles away, who is incapable of helping you.

In a firefight, he who hesitates is lost – usually forever.

In a firefight, delay is the deadliest form of denial.

On the keyboard of combat, keep one finger on the *escape* key.

On the keyboard of combat, keep one finger on the *escape* key.

Always know when to "get out of Dodge."

In a firefight, (1) one problem is a problem, (2) two problems means it's time to get out of Dodge, but (3) three problems often means it's *too late* to get out of Dodge.

In combat, "what" is always more important than "why" (when you see a snake, don't fret over why it's there, just shoot it).

A *good plan* for today is better than a *great plan* for tomorrow.

In combat, if something doesn't matter, it doesn't matter. The trick is to *make sure* it doesn't matter.

A *good plan* for today is better than a *great plan* for tomorrow.

On patrol and ambush, smart warriors always keep their eyes peeled, for he who sees first, lives longest.

If you lose contact with the enemy, remember to look behind you.

Anything you do in combat can get you killed. Doing nothing usually will get you killed more quickly.

In a firefight, do *something*, even if it's wrong.

No matter how bad it gets, it's not over until it's over.

If you lose, and if you're still alive, don't lose the reason.

– Murphy's Infantry Precautions –

Beware! Combat is always easier to get into than out of.

Remember, you aren't Superman, and you aren't bulletproof.

In combat, the best medal is the longevity medal.

Once you're in the fight it's too late to pause and ponder whether or not it was a good idea.

Combat isn't like Hollywood. In combat, getting shot *hurts*.

If you're in it, there's no such thing as a *little* firefight.

After a firefight, the best medal is the longevity medal.

In a firefight, prayer may not help – but it certainly can't hurt.

In a firefight, prayer may not help – but it certainly can't hurt.

"Courage under fire" means that you're the only person who knows that you're scared.

In a fight-to-the-death, a *tie* or a *split decision* isn't a viable option.

When you've run out of options in a firefight, if you think the enemy may be low on ammo, try to look unimportant.

Never try to draw fire. It irritates those around you.

In hi-tech, bio-weapon, and smart-weapon combat, there's no safety in numbers – or in anything else.

Avoid all loud noises. There are few silent killers in combat.

Remember, the bad guys may surrender, but their mines won't.

There are good plans, but there are no perfect plans. Beware, for your confidence may be your suspicion, asleep.

Never try to draw fire. It irritates those around you.

In combat, if everything is as clear as a bell and things are going as planned, look out! You're ripe for a surprise.

A warrior who thinks *small bad guys* can't be lethal has never been in bed with a *small rattlesnake*.

In combat a good plan is a good idea, but never get married to it. No plan ever survives enemy contact, intact.

When you're curled up deep in your hole with mortars and artillery and bombs exploding all around, you can bet your bottom dollar that no (1) pacifists, (2) liberals, or (3) atheists are lurking nearby.

A warrior who thinks *small bad guys* can't be lethal has never been in bed with a *small rattlesnake.*

When you have a clear choice in a crisis, opt for safety. That way, you'll survive to be brave later on.

When your situation is desperate, it's too late to be serious.

In combat, all macho talk aside, *surviving* is more important than *winning.* You can only die once. But, you'll have many chances to win *if you survive* long enough to get them.

He who hesitates under fire usually won't get another chance to.

When you're curled up deep in your hole, with mortars, artillery, bombs, and missiles exploding all around, you can bet your bottom dollar that no (1) pacifists, (2) liberals, (3) atheists, (4) staff pogues, (5) attorneys, or (6) brain-dead advocates of political correctness are lurking nearby.

When the pin has been pulled, Mr. Grenade is no longer our friend.

In combat, if you find yourself in front of your fellow warriors, they likely know something you don't.

In combat, rash decisions often *bleed* consequences.

Any idiot can charge an enemy machinegun, across open terrain, alone, in broad daylight – once.

When the pin has been pulled, Mr. Grenade is no longer our friend.

Remember, the *killing range* of any grenade is always much greater than your *jumping range*.

In combat it's a bad idea to buddy-up with anyone whose grenade *throwing range* is less than the grenade *killing range*.

In a firefight, if you're keeping your head while all around you are losing theirs, perhaps it's time for a reality check.

The *killing range* of any grenade is always much greater than your *jumping range*.

Food for thought: your weapon was made by the lowest bidder.

True, a sucking chest wound is bad. But, (1) all wounds are bad, and (2) all wounds suck.

A non-posthumous Purple Heart merely proves you were (1) smart enough to think of a plan, (2) crazy enough to try it, and (3) lucky enough to survive.

Getting outnumbered and surrounded by the enemy usually is a bad idea. But, on the bright side, it's a great time to get rid of the heavy extra ammo you've been humping.

Before any firefight, it's bad luck to be superstitious.

Any bad guy with a *rifle* is a better shot than you with a *pistol*.

In a firefight, the greatest danger is the company of scared people.

Yet, a good *scare* is usually more effective than good *advice*.

In a firefight, the greatest danger is the company of scared people.

If your attack is known in advance, it should not take place.

After the fray, it's better to be a *live lamb* than a *dead lion*.

To appreciate the value of a single minute, ask any combat veteran what his buddy was doing a minute before he got killed.

Helicopter resupply pilots will see you, but attack pilots on bombing or strafing runs will not – dig your hole a little deeper.

– Murphy's Infantry Ironies –

No matter what you do, the bullet with your name on it will get you. Also, so can *random bullets* addressed "to whom it may concern" and *shrapnel* addressed to "occupant."

For each military action there will be an equal and opposite criticism.

In a firefight when you have plenty of ammo, you never miss. When you're low on ammo, you can't hit diddly-squat.

The ammo you need *now* will be on the *next* chopper.

The ammo you need *now* will be on the *next* chopper.

If you wear body armor, the bad guys usually will miss that part.

Combat, like love, isn't called off on account of darkness.

In combat the only thing more accurate than incoming *enemy fire* is incoming *friendly fire*.

Most firefights occur at the junction of several maps. If you don't use paper maps, firefights occur when your batteries die.

If you can't remember, the Claymore is pointed toward you.

Standard five-second fuses burn in five seconds – plus or minus four.

If you can't remember, the Claymore is pointed toward you.

No *combat-ready* unit ever passed inspection. No *inspection-ready* unit ever passed combat.

It's impossible to make any weapon system foolproof, because fools are deceptively ingenious.

In any wooded area at night, the sharp dead limbs on trees will be at (1) eye level, or (2) groin level.

Professional enemy soldiers are predictable. Unfortunately the world is full of dangerous amateurs.

In any wooded area at night, the sharp dead limbs on trees will be at (1) eye level, or (2) groin level.

Combat isn't like Hollywood. In combat the cavalry doesn't always come charging to your rescue.

Remember, if the enemy is within range, so are you.

Remember, if the enemy is within range, so are you.

Radar fails (1) at night, or (2) in inclement weather, or (3) both.

Radios fail when you desperately need fire support.

If you make it hard for the enemy to get in, you can't get out.

The enemy always mines the *easy* way out.

The easy way usually will get you killed.

Often, *military intelligence* is a contradiction in terms.

The enemy diversion you ignore is often his main attack.

The enemy will attack when two conditions are met: (1) when he is ready, and (2) when you are not.

The enemy will attack when two conditions are met: (1) when he is ready, and (2) when you are not.

The enemy will attack *most ferociously* under two conditions: (1) on the darkest night, and (2) during a rainstorm.

Fortify your front, and the enemy will attack your rear.

If your ambush is properly set, the enemy will never arrive.

If you're sufficiently dug-in, the enemy will never attack.

Superior firepower *always* prevails – sometimes.

The enemy has a penchant for converting your mines into equal opportunity weapons.

The enemy has a penchant for converting your mines into equal opportunity weapons.

Combat *experience* is what we call our combat *mistakes*.

Combat *experience* is what we call our combat *mistakes*.

Warriors who take more than their fair share of objectives will get more than their fair share to take.

If it worked in practice, it will fail in combat.

– Murphy's Military Logic –

Interchangeable parts usually ***aren't***.

Friendly fire always ***isn't***.

Insect repellants often ***don't***.

Perfect plans ***aren't*** (and neither is the back-up plan).

Recoilless rifles ***aren't*** (ask the idiot who stood behind one).

Waterproof clothing ***isn't*** (but it *will* retain perspiration).

Flash suppressors ***do*** (but only in the daytime).

Suppressive fires ***do*** (but only on abandoned positions).

Things that must be shipped together usually ***aren't***.

Things that must work together, for some reason, ***won't***.

Things that must work together can't be shipped together.

Whenever you need ***n*** crucial items, there will be ***n-1*** in stock.

In combat, no order is so simple that it can't be misunderstood.

Experience in combat is something you never have enough of until *after* you need it.

In a firefight, teamwork is essential. It gives the enemy someone else to shoot at.

In combat, the side with the simplest uniform usually wins.

In a firefight, teamwork is essential. It gives the enemy someone else to shoot at.

The thing you need the most will be at the bottom of your pack.

If your planned attack looks easy, it will be *hard.* If your planned attack looks hard, it will be *impossible.*

Loose pins can cause your grenades to detonate. This will make you rather unpopular with what's left of your unit.

Unfortunately, tracers work both ways.

In military intelligence, (1) the information you have isn't what you want, (2) the information you want isn't what you need, (3) the information you need isn't available, and (4) everything depends upon something else.

No matter which way you march, it's uphill and into the wind.

In combat, the side with the simplest uniform usually wins.

You always will be downwind when CS gas is used.

Any stone in your boot migrates to the point of maximum pressure.

The weight of your pack will never remain uniformly distributed on the shoulder straps.

When hot garrison chow is flown to the field, it will rain.

About 15 percent of an intelligence report will be accurate. The trick is to know *which* 15 percent.

Highly trained military working dogs are trained to attack anybody, including you.

In military intelligence, (1) the information you have isn't what you want, (2) the information you want isn't what you need, (3) the information you need isn't available, and (4) everything depends upon something else.

– Murphy's Barracks Wisdom –

No warrior fights all the time. Between battles and wars, he and his warrior brothers will retire in-the-rear-with-the-gear to rest and recuperate. There they will carouse, drink, brawl, and attempt to corrupt members of the opposite sex.

Warriors, your days in a barracks environment are fraught with peril. Here the *new enemy* smiles and shakes hands before stabbing you in the back. Soulless media whores, the mercenary purveyors of sensationalism and negativism, will leap at any opportunity to steal your honor. Worse yet, brain-dead *politically correct* staff pogues will stop at nothing in their attempts to drag you into the gutter of society with them. Therefore, all warriors in a barracks environment should heed ***Murphy's Barracks Wisdom***:

Brain-dead *politically correct* staff pogues will stop at nothing in their attempts to drag you into the gutter of society with them.

Between wars, warriors should never delay (1) the end of a meeting, or (2) the start of Happy Hour.

Between wars, never let a fool kiss you, or a kiss fool you.

Between wars, *girlfriends* are fair game, but *wives* are not.

Between wars, *girlfriends* are fair game, but *wives* are not.

Remember, when media whores can choose between a pitiful little cry-baby and a true warrior, they always interview the cry-baby.

Those who torture animals and wet the bed are either sex perverts or staff pogues. Both should be avoided.

Friends come and go, but enemies *accumulate*.

Good advice: "You have the right to remain silent."

No decision, made in your absence, will be in your best interests.

> When media whores can choose between a pitiful little cry-baby and a true warrior, they always interview the cry-baby.

Between wars and battles, while carousing and brawling in-the-rear-with-the-gear, warriors may be accused of some infraction of civil or military protocol. This is an occupational hazard, but Murphy can enable a warrior to avoid blame for most such infractions. When accused of some dastardly deed, a wise professional warrior will start at the top of ***Murphy's Responses to Avoid Blame*** and work his way down – but only as far as necessary:

– Murphy's Responses – to Avoid Blame

1. Who, me?
2. I wasn't there!
3. I didn't do it!
4. Nobody saw me do it!
5. You can't prove a thing!
6. That's my story, and I'm sticking to it!

Murphy's Military Definitions for Infantry

Listen-up! These military definitions are crucial need-to-know information for the warrior elite, the ***Magnificent Grunts***. Do you recall the true meaning of *Cranial-Rectal Inversion Syndrome*? How about *scrounge*? Do you know the definition of *marriage*? What is a *Mail Buoy Watch*? All the proper answers are in this chapter. They apply to all warriors, but they're essential for the world's premier combatants, the Grunts.

In a separate chapter, Murphy shares his knowledge of *Aviation* definitions. All Grunts should take a quick look at them, just to be on the safe side. However, Murphy cautions Grunts not to dwell on the aeronautical perils for very long. Those who get hung-up on aviation hazards will *never* agree to ride in any type of aircraft again.

Do you recall the true meaning of *Cranial-Rectal Inversion Syndrome*?

Detailed below are ***Murphy's Military Definitions for Infantry***. All warriors, and especially the Grunts, should read and heed:

Aircraft: (1) The precursor of the Frisbee. (2) The generic name for any noisy heavier-than-air flying contraption. (3) The most unreliable and perilous mode of travel on Earth.

Aircraft Carrier: The biggest, the most expensive, the slowest-moving, most explosive-filled bull's-eye on Earth.

Ballistic: A characteristic of First Sergeants and Sergeants Major.

ALICE: (1) In *theory*, a good-time companion on leave or liberty. (2) In *reality*, not much fun in the deep end of the pool.

Ammo: (1) For warriors, a possession more valuable than gold. (2) The international "currency" for the 21st Century.

– Murphy's Guide – to American Warriors

Airman: (1) A flaccid civilian detainee in the U.S. Air Force. (2) A pitiful sniveling back-stabbing, donut-eating, risk-avoiding, public assistance program reject. (3) A useless lethargic drone whose lack of initiative, intellect, and stamina render him incapable of finding employment in the private sector.

Marine: (1) An unfortunate person who lacked the brains to get in the Air Force, the skills to get in the Navy, and the common sense to settle for the Army. (2) A member of the group of outcasts known as *Uncle Sam's Misguided Children.* (3) A person for whom all reading and writing test requirements have been waived. (4) A person for whom "Ooorah!" is a proper response to all questions from a superior.

Sailor: (1) A primitive pre-Neanderthal life form. (2) A deck ape. (3) A nautical paint-picker. (4) A member of the underclass of dark and slimy squid-like creatures, most of whom have been banished to their natural habitat in the murky depths of the ocean where normal human beings don't have to associate with them.

Soldier: (1) A member of the world's largest civil bureaucracy. (2) A person easily recognized by (a) his trousers, which are too short; (b) his hat, which is too large; (c) the huge pockets on his trousers, in which he can warm his dainty hands; and (d) the vast assortment of emblems, crests, badges, and shiny doo-dads that adorn his uniform, which is similar in appearance to that of a Greyhound bus driver.

Ammo Dump: A place a wise warrior will avoid during incoming mortar, rocket, or artillery fire – even if it is a deep hole.

Armored Vehicle: The generic name for mobile steel conveyances that attract armor-piercing incendiary rounds.

A fully automated mobile cannon from Future Combat Systems sends a high explosive round down-range (photo courtesy of U.S. Army).

Artillery: A weapon designed to kill or maim as many of the enemy and his evil cohorts as possible and restore cave-dwelling as an acceptable way of life in the former enemy territory.

Ashore: The nebulous Navy term that means anywhere in the air, on land, or at sea – except on base.

Artillery: A weapon designed to kill or maim as many of the enemy and his evil cohorts as possible and restore cave-dwelling as an acceptable way of life in the former enemy territory.

Ballistic: A characteristic of First Sergeants and Sergeants Major.

Bayonet: (1) A weapon of last resort. (2) A poor primary-weapon selection for use in a firefight.

Bayonet Fighting Expert: A wise and experienced warrior who knows the Vertical Buttstroke isn't a sexual technique.

Beachhead: The guaranteed bull's-eye for enemy mortars, artillery fire, and aerial bombardment.

BLT: A mechanized amphibious force of highly skilled assassins.

Bravery: In common usage, usually a synonym for *sheer stupidity.*

Board of Inquiry: A devious and spurious group of staff pogues who, given enough time and data, can prove anything.

Bravery: In common usage, usually a synonym for *sheer stupidity.*

Brig (or Stockade): (1) A poor selection for military lodging, even though the *rooms* are free-of-charge. (2) For all macho warriors, an even worse choice if the rest of their unit is in combat.

Brown-Bagger: (1) A well-intentioned warrior who has stumbled into the dark and evil snake-pit of marriage. (2) A term derived from the small *brown paper lunch bag* carried by married warriors, who can no longer afford to buy their own meals.

Carry On: A verbal order which means, *resume doing nothing.*

Chinese Fire Drill: Any military endeavor noteworthy for its absence of coordination and purpose.

Cinderella Liberty: A devious staff pogue ruse designed to promote sobriety among macho warriors.

Close: A military near-miss, a matter of utmost concern in the related arts of (1) horseshoe throwing, and (2) grenade tossing.

Brown Bagger: (1) A well-intentioned warrior who has stumbled into the dark and evil snake-pit of marriage. (2) A term derived from the small *brown paper lunch bag* carried by married warriors, who can no longer afford to buy their own meals.

Close Air Support:
A good thing to have – if it's not *too* close.

College Campus: An academic retreat where sanctimonious apostles of *political correctness* outnumber men of principle and honor.

Carry On: A verbal order which means, *resume doing nothing.*

Combat: What a warrior is in when he has run out of everything except the bad guys.

Combat Breakfast: Two aspirins, two cups of coffee (if available), a quick prayer, and a quick puke.

Combat Experience: The sum of your *combat mistakes.*

Combat Experience: The sum of your *combat mistakes.*

Combat Pay: (1) A flawed concept. (2) A premise which allows the government to *save money* by temporarily paying a warrior *more money* in anticipation of his expedited demise.

Communism: (1) An ideology embraced by Marx, Engels, and other ignoramuses. (2) A form of socialism designed to impoverish any governmental entity. (3) A concept that would work only in Heaven, where it isn't needed, and in Hell, where it's already established. (4) An easy way to raise suffering to a higher level.

Conclusion: What you will reach when you get tired of thinking.

Computer: (1) An electro-mechanical marvel, the operation of which is beyond the intelligence level of the intended military user. (2) The primary cause of profanity among warriors.

Conclusion: What you will reach when you get tired of thinking.

Corpsman (or Medic): In combat, a good guy to buddy-up with.

Cranial-Rectal Inversion Syndrome: (1) A debilitating staff pogue illness. (2) A progressive and degenerative malaise. (3) A chronic mental paralysis believed, by medical experts, to be the result of an in-the-rear-with-the-gear practice too uncouth to describe.

Critical Terrain: Terrain which, if not grabbed or camped out upon, will make a warrior the *screwee* in offensive or defensive warfare.

D-Day: The day before which a warrior's insurance documents and Last Will and Testament should be completed.

Death Before Dishonor: The most popular tattoo among warriors.

Combat Pay: A flawed concept.

Defilade: When the shooting starts, the best position to be in.

Demilitarized Zone: A zone that should be, but usually isn't.

Diplomacy: The art of explaining to the bad guys the manner and certainty of their impending demise if they refuse to surrender.

Fool: A warrior who still believes in fighting fair.

Drill Instructor (or Drill Sergeant): A maniacal, sadistic, extremist psychopath whose name a warrior will never forget.

Double-Tap: The professional warrior's policy of firing two rounds into the head of each dead bad guy – *just to make sure*.

Field Day: A despised indoor non-athletic activity.

Double Tap: The professional warrior's policy of firing two rounds into the head of each dead bad guy – *just to make sure*.

Fighting Hole: (1) Formerly called a *foxhole*. (2) The technical name for a small military *hiding hole*.

Flex: A really cool sounding and non-doctrinal term to explain how your unit will maneuver, under fire, from one battlefield location to another, when no one has a clue.

Fool: A warrior who still believes in fighting fair.

Forward Air Controller: (1) A "FAC," pronounced *fack.* (2) A warrior, trained in aerial bombardment techniques, who has been assigned to a company or battalion of Grunts. (3) A person to be watched at all times (warriors interested in longevity should dive into a deep hole the instant their FAC begins chatting on his radio, for smoke and loud noise often follows).

Four-Wheel-Drive: A motor vehicle capability that enables warriors to get stuck in the mud at more remote and inaccessible locations.

Geneva Convention: A symposium of civilians who made up silly rules that warriors must follow – unless no one is watching.

Good Judgement: Sound military decision-making ability derived from *experience*, which is derived from *bad judgement*.

Grunt: (1) A professional assassin. (2) An infantryman. (3) The sole reason for the existence of helicopters. (4) An indispensable element of the foreign policy of the United States. (5) An elite and feared warrior who will go anywhere, at any time, and destroy whatever, or whomever, he is ordered to destroy – as long as he is allowed to sing obscene songs, kick cats, drink, brawl, embellish War Stories beyond recognition, and corrupt members of the opposite sex.

Heroism: A trait often displayed in combat after a warrior has run out of all other viable options.

Gulf War: The common name for four days of intense target practice in the land-of-sand in 1991.

Gungy: An up-beat, enlightened, euphoric, gung-ho state of mind.

Hero: An illustrious term that all red-blooded warriors apply to themselves – *in the club after the fourth drink.*

H-Hour: The most introspective time of a warrior's day.

Get some with the M-240G (photo courtesy of U.S. Marine Corps).

Head (or Latrine): In times of utter confusion, the small room in which a warrior *still* should know what he's doing.

Hero: An illustrious military term that all red-blooded warriors apply to themselves – *in the club after the fourth drink.*

H-Hour: The most introspective time of a warrior's day.

Heroism: A trait often displayed in combat after a warrior has run out of all other viable options.

Hey-diddle-diddle: (1) With reference to a proposed assault, words which describe an absence of analytical thought. (2) An offensive warfare tactic that guarantees an 80 percent or higher casualty rate.

HMMWV, Humvee, Hummer, or whatever: The western world's most expensive four-wheel-drive military or civilian play-toy.

Journalist: A slovenly brain-dead media whore who meticulously separates the wheat from the chaff – and publishes the chaff.

LAV: (1) A small motorized conveyance with an identity crisis. (2) A younger brother to a tank. (3) A thin-skinned steel box primarily useful as a crematorium for its talented occupants.

LCAC: (1) A huge 50 knot flying carpet. (2) A magical amphibious machine that can defy gravity and all known laws of physics.

LHA: The official military acronym for Luxury Hotel Afloat.

M-2 Bradley Fighting Vehicle: (1) An uncomfortable and unreliable conveyance. (2) A dangerous metal box (think, *target*) that warriors should avoid like bubonic plague during a tank battle.

LHA: The official military acronym for Luxury Hotel Afloat.

M-9 Service Pistol: A small handgun which, in a firefight, a warrior should exchange for a large automatic weapon.

M-16A2 Rifle: A 5.56mm magazine fed, gas operated, air cooled, shoulder fired weapon – manufactured by the *lowest bidder*.

M-18A1 Claymore Mine: A lethal directional antipersonnel mine upon the side of which – after deployment in combat – warriors hope they can't read the words, "Front, Toward Enemy."

M-203 Grenade Launcher: The deadly breech loaded, single shot, shoulder fired, mini-bomb launcher, for which a tactical nuclear round would be an excellent idea.

M-9 Service Pistol: A small handgun which, in a firefight, a warrior should exchange for a large automatic weapon.

M-240G Machine Gun: A heavy 7.6mm automatic weapon that you hope some warrior, *other than yourself*, has to carry.

Maggie's Drawers: (1) For all red-blooded warriors, the only set of red drawers they *never* want to see. (2) An embarrassing insult.

Mail Buoy Watch: (1) A crucial nocturnal assignment for all new and useless U.S. Navy ensigns. (2) An aquatic snipe hunt.

Maggie's Drawers: For all red-blooded warriors, the only set of red drawers they *never* want to see.

Map: An archaic land-navigation aid which is totally useless on the modern-day battlefield – unless all your batteries have died.

Marriage: (1) Not a military term. (2) A civil snake-pit into which well-intentioned warriors sometimes stumble. (3) The tragic result of trying to *think* with the wrong part of one's anatomy. (4) A progressive three-ring circus: engagement ring, wedding ring, and suffering. (5) The number one cause of divorce.

Medal: (1) The generic name for any one of a multitude of eye-catching uniform trinkets. (2) The shiny doo-dads that all warriors want more of – *exclusive* of the Purple Heart.

Medal: The shiny doo-dads that all warriors want more of – *exclusive* of the Purple Heart.

MEU: Next to a Class 5 hurricane or a 700 foot high tsunami, the most lethal force on planet Earth.

MEU (SOC): A lethal force *even more* destructive than a Class 5 hurricane or a 700 foot high tsunami.

Mines: The generic military term for Equal Opportunity Weapons.

Mortar: The heavy bulky weapon that warriors least like to carry.

MOUT: (1) Exclusive of the north pole and the south pole, the last environment on Earth where wise warriors want to fight. (2) A place where discretion *really* is the better part of valor.

MPC: (1) Funny-money. (2) As worthless as Monopoly money.

Napalm: (1) Incindi-gel. (2) An excellent area support weapon.

Nomex: A synthetic fiber known for its penchant to burst into flame at moderate temperatures and be impossible to extinguish.

Napalm: Incindi-gel. An excellent area support weapon.

Ode to Naivety (posthumous): "Often wrong, but never in doubt."

Old Corps, The: (1) The era of the M-1 Garand. (2) The era when John Wayne was still on active duty. (3) The era when the Chinese Army didn't have enough rowboats to invade Taiwan. (4) Or, any bygone era before *any* of the above.

Payback: An inspirational experience – if a warrior survives it.

Peace Is Our Profession: The worst military motto ever dreamed up by staff pogues (true warriors prefer Gen. Curtis E. LeMay's credo: "We'll bomb them back into the Stone Age").

Perimeter Defense: What Gen. George A. Custer, USA, and his Seventh Cavalry should have established at the Little Big Horn on 25 June 1876, shortly before their untimely demise.

Perimeter Defense: What Gen. George A. Custer, USA, and his Seventh Cavalry should have established at the Little Big Horn on 25 June 1876, shortly before their untimely demise.

Purple Heart: The *least desirable* medal awarded to warriors in the U.S. Armed Forces.

PFM: The military acronym for the comprehensive easy-to-understand non-technical explanation for why complex military equipment functioned as it did – when no one has a clue.

Pogey Bait: (1) The most nutritious of the four major food groups. (2) The main energy source of choice for experienced warriors, especially when their cholesterol level is too low.

Political Correctness Counselor: (1) A civilian advisor who is depriving a village, somewhere, of its idiot. (2) A brain-dead person who would be out of his depth in a parking lot puddle. (3) An ignoramus. (4) A hand-wringing sniveling idiot.

Preparation Fire: (1) Commonly called *prep-fire*. (2) Smoke and loud noise created to promote unjustified confidence among members of a waiting assault force.

Professional Reading List: A list of inspirational military books that ***will never include*** (1) *Fighter Aces of the Iraqi Air Force* or (2) *Victory the McNamara Way*.

Radio: When in a fire-fight, a useless electronic suggestion box.

Purple Heart: The *least desirable* medal awarded to warriors in the U.S. Armed Forces.

R & R: In *theory*, the military acronym for "Rest and Recuperation" far from the combat zone (however, in *reality*, the proper acronym should be "I & I").

Radio: When in a firefight, a useless electronic suggestion box.

Rate of Fire: (1) The *theoretical* number of rounds an automatic weapon can fire in one minute. (2) A *meaningless term*, because the weapon would melt if any idiot ever tried it.

Pogey Bait: The most nutritious of the four major food groups.

Recon (or Special Forces, or SEAL): (1) The *stealth* version of the basic U.S. Marine, U.S. Army soldier, or U.S. Navy sailor. (2) A warpainted assassin who prefers to fight with his knife, bayonet, teeth, and fists – just to conserve ammo.

Reconnaissance by Fire: A noisy technique used on patrol when a warrior can't see through the thick bushes.

REMF (spoken, *ree-miff*): The official military acronym for Rear-Echelon-Mother-Fu[expletive]. The derogatory term used by combat infantrymen and aircrews to describe military personnel in safe administrative and support roles, far from the fighting – even though such persons are sometimes secretly envied.

Reconnaissance by Fire:
A noisy technique used on patrol when a warrior can't see through the thick bushes.

Roach Coach: A civilian-owned vehicle stocked with quasi-edible products designed to induce diarrhea, heartburn, and other symptoms of acute gastrointestinal distress.

Rug Dance: (1) A warrior's rhythmic shuffling endeavor that requires no partner. (2) A spirited form of dancing during an extraordinarily one-sided chat with a warrior's irate superior.

Rules of Engagement: The childish rules that warriors must *claim* they followed to the letter – as long as they have killed all the bad guys and witnesses who might have claimed otherwise.

Smoking Lamp: A nonexistent lamp that can, nonetheless, be lit.

Sniper's Motto:
"Reach out and touch someone."

Scrounge: (1) A *verb*; meaning to obtain by a devious method. (2) A *noun*; meaning a warrior who is adept at obtaining, by devious methods, anything from crates of frozen steaks to tactical atomic weapons. (3) A comshaw artist.

Short-Arm Inspection: A former practice rendered obsolete by the pharmaceutical miracles of modern medicine.

Stuff: A nebulous military term that refers to (1) a tangible thing, or to (2) a situation, condition, or process, as exemplified below:

This is rough stuff: Typical statement of an ***Air Force NCO*** while driving his air-conditioned sedan two miles from his air-conditioned office to his air-conditioned hotel, ***in the rain.***

This is really rough stuff: Typical statement of an ***Army Ranger***, weapon at sling arms, humping a 40 pound pack, after jumping from an aircraft, marching 10 miles to an enemy village, and waiting for reinforcements, ***in the rain***.

This is horrible stuff: Typical statement of a ***Navy SEAL***, lying in the mud with his 60 pound pack, weapon in hand, after jumping from an aircraft, swimming a mile to shore, crawling 15 miles to an enemy village, and preparing to attack, ***in the rain***.

I love this stuff: Typical statement of a ***Marine Recon***, up to his eyeballs in a vermin-infested swamp with his 80 pound pack, a weapon in each hand, after jumping from an aircraft, swimming two miles to shore, slithering 20 miles through a swamp, killing several alligators while negotiating the swamp, assaulting the enemy village, killing all the bad guys and witnesses, and slithering back into the slime of the swamp in order to ambush any more bad guys and witnesses who might wander by, ***in the rain***.

Slopchute: (1) A geedunk that serves libations. (2) A facility not noted for stocking nutrients of the four major food groups.

Smoking Lamp: A nonexistent lamp that can, nonetheless, be lit.

SNAFU / FUBAR / SAPFU: Military acronyms listed in ascending order of implied ineptitude and/or stupidity.

Snake and Nape: In a protracted battle, a wonderful thing to find that your aviation brothers-in-arms have plenty of.

Sniper's Motto: "Reach out and touch someone."

Straight Scoop: Factual information, as opposed to a new PFC or Airman or Seaman who reports, "I just got the word . . . ."

Snake and Nape: A wonderful thing to find that your aerial brothers-in-arms have plenty of.

Sucking Chest Wound: Nature's way of telling you to slow down.

Supporting Fire: An excellent thing to witness – if it's yours.

Surrender: Dastardly conduct that wimps and girlie-men engage in, *especially* if they are into (1) masochism or (2) cold rice-balls.

Tank: (1) A tracked steel box crammed full of explosives. (2) An efficient crematorium for its highly skilled crew. (3) Otherwise, a monument to the inaccuracy of direct fire.

Sucking Chest Wound: Nature's way of telling you to slow down.

Target of Opportunity: A bad thing to become, or be mistaken for.

Terrorist: A freedom fighter with a different perspective.

TRICARE: A *sick joke* on a warrior and his family.

Tank: An efficient crematorium for its highly skilled crew.

Vampire: The common name for either a (1) this-is-no-drill enemy weapon, or a (2) nocturnal Transylvanian blood-sucker.

Vietnam: A tropical place where it *did too* get cold at night.

Vietnam War: A faraway armed conflict pitting the principles of Thomas Jefferson against the principles of Karl Marx.

War: (1) In general, the unfolding of miscalculations. (2) A long series of armed conflicts during which hundreds of American Warriors earned medals for valor in combat every single day, and during which – from time to time – a few were actually awarded.

Warrior: (1) A street-legal government assassin. (2) A highly skilled population control specialist. (3) A romping, stomping, devil-may-care purveyor of death and destruction. (4) An overbearing psychopath. (5) A killer by day, lover by night, and drunkard by choice. (6) One who can curse for ten minutes without repeating a word. (7) One who dreams of using media whores and liberals for target practice. (8) A red-blooded patriot who knows that "Kill, sir!" is the proper response to all questions. (9) One who knows that "sodomy" and "politically correct" should fall into the same UCMJ sub-chapter.

– Murphy's Abridged Guide – to the U.S. Armed Forces

U.S. Air Force: (1) A government funded ***flying club***. (2) An organization consisting of prima donna wild-blue-yonder wannabes plus a host of brain-dead sluggards and whiny-babies who have been unable to find gainful employment in the private sector.

U.S. Army: (1) A government funded ***gun club***. (2) A bureaucracy that provides free lodging and remedial training to prison parolees and deranged Rambo wannabes who have been unable to cope with routine day-to-day pressures of society.

U.S. Marine Corps: (1) A government funded ***penal institution***. (2) A rehabilitation program devoted to the welfare of a motley collection of incorrigible psychopaths, misfits, and confused slow-learners known as Jarheads (whatever *Jarheads* may be).

U.S. Navy: (1) A government funded ***cruise service***. (2) The most effective way to quarantine the slovenly underclass of mentally degenerate squid-like creatures far out at sea, where they will be unable to mate and cause further damage to the world's gene pool.

Murphy's Laws of Combat for Aviation

The history of military aviation is an immense sea of errors in which a few obscure truths may here and there be found. And, like other occult techniques, aviation has a private jargon contrived to obscure its methods from non-practitioners.

Flying without feathers was *never* easy.

Flying without feathers was *never* easy. First came the hot air balloonists. They never killed anyone except themselves, from time to time. But on 17 December 1903 two brothers, Wilbur and Orville, started the aerial foolishness that still gets warriors in trouble today. Because of those two culprits, aviation has gotten ever higher, ever faster, ever more complex, ever more dangerous.

Today flying machines of incredible lethality can swoop down into combat. For pilots and aircrewmen, the trick is to stay alive long enough to get the needed experience to enable them to stay alive a little bit longer. You Grunts along for the ride, *think* about it. Maybe humping the hills on foot isn't so bad after all.

For pilots and aircrewmen, the trick is to stay alive long enough to get the needed experience to enable them to stay alive a little bit longer.

In military aviation we find that the most deadly enemy is not the bad guys. The biggest threat comes from our flimsy flying contraptions that *screw* (helicopters) or *suck and blow* (jet planes) their way through the sky. To assist warriors who fly, Murphy offers his unique aeronautical knowledge in four categories:

1. Murphy's Aviation Philosophy
2. Murphy's Aviation Admonitions
3. Murphy's Aviation Ironies
4. Murphy's Aviation Predictions

– Murphy's Aviation Philosophy –

Gravity never loses. The best a warrior can hope for is a draw.

Although fuel is a limited resource, gravity is forever.

In the ongoing battle between (1) military aircraft going hundreds of miles per hour, and (2) the mountains going zero miles per hour, the mountains have yet to lose.

The three military aviation constants: (1) airspeed is *life*, (2) altitude is *life insurance*, and (3) fuel is *more* life insurance.

> The mechanics of military flying are simple. If the pilot pushes the stick forward, the houses on the ground look bigger. If he pulls the stick back, the houses look smaller. However, if he pulls the stick back *too far* the houses *very rapidly* look bigger again.

The only time your aircraft has too much fuel is when it's on fire.

A combat aircrew lives in a world of perfection – or not at all.

If all a military pilot can see out of the cockpit window is the ground, going around and around, and all he can hear is screaming back in the cabin, something likely is amiss.

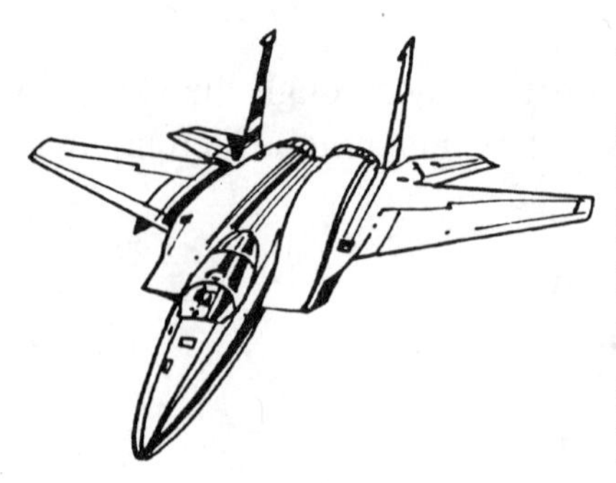

The mechanics of flying are simple. If the pilot pushes the stick forward, the houses on the ground look bigger. If he pulls the stick back, the houses look smaller. However, if he pulls the stick back *too far* the houses *very rapidly* look bigger again.

Assumption is the mother of most crashes.

Military flying consists of hours and hours of boredom, interrupted by brief moments of stark terror.

All takeoffs are optional. But landing, *somewhere*, is mandatory.

A military pilot can land anywhere – once.

While flying, you usually don't know what you don't know.

When returning to Earth at high speed, the probability of survival is *inversely* proportional to the angle of arrival (large angle of arrival, small probability of survival, and vice versa).

While flying, you usually don't know what you don't know.

In combat, having a wingman is essential. He gives the bad guys somebody else to shoot at.

When flying, it's much better to (1) *look bad* than to (2) *die*. But in military aviation it's easy to do both simultaneously.

There are (1) *old* combat aircrews, and there are (2) *bold* combat aircrews. But there are very few *old and bold* combat aircrews.

Aerial combat is the perfect vocation for warriors who want to feel like young boys – but not for warriors who still are.

There's no such thing as a *routine* combat mission.

Superior pilots and aircrewmen use their *superior judgement* to avoid situations where they might have to use their *superior skills*.

If you don't know who the world's greatest pilot (or crew chief, or gunner, or loadmaster) is, he isn't you.

The late great SR-71 Blackbird (photo courtesy of U.S. Air Force).

While on your takeoff roll, if an earthquake opens a 100 foot chasm across the runway, and you crash into it, the mission of the Accident Board will be to find a way to blame it on pilot error.

Asking a military pilot what he thinks about the FAA is like asking a dog what he thinks about fire hydrants.

Asking a military pilot what he thinks about the FAA is like asking a dog what he thinks about fire hydrants.

A twin-engine aircraft doubles your chance of an engine failure. And, after one engine has failed, the most common purpose of your other engine is to fly you to the scene of your accident.

Aerial combat is the perfect vocation for warriors who want to feel like young boys – but not for warriors who still are.

Any tactic, done twice without crashing, becomes a procedure.

There are aviation *rules*, and there are aviation *laws*. The rules were made by men, and can be suspended. The laws were made by the Deity, and should not be trifled with.

A twin-engine aircraft doubles your chance of an engine failure.

A military pilot who relies on a "terminal forecast" can be sold the Brooklyn Bridge. If he also relies on a "winds aloft" report, he can be sold Niagara Falls too.

Flying at night is almost as easy as flying in the daytime, because the airplane doesn't know it's dark.

Any tactic, done twice without crashing, becomes a procedure.

You can get *anywhere* in ten minutes if you fly fast enough.

But, you've *never really* been lost until you're lost at Mach 2.

The two worst things that can happen to an old military aviator:

1. One sunny day he will take off and soar skyward, ***knowing*** it will be his last flight.
2. One sunny day he will take off and soar skyward, ***not knowing*** it will be his last flight.

– Murphy's Aviation Admonitions –

When in doubt, climb! No pilot has ever collided with the sky.

Remember, you can only *tie* the record for flying low.

If you absolutely must fly *low*, don't fly *slow*.

Military pilots and aircrewmen who hoot with the owls by night must not try to soar with the eagles by day.

No matter how bad it gets, *fly the aircraft!* Fly it until the last piece stops moving. Remember: (1) aviate, (2) navigate, (3) communicate.

Don't crash while trying to fly the radio. Military aircraft fly because of the principle discovered by Bernoulli, not Marconi.

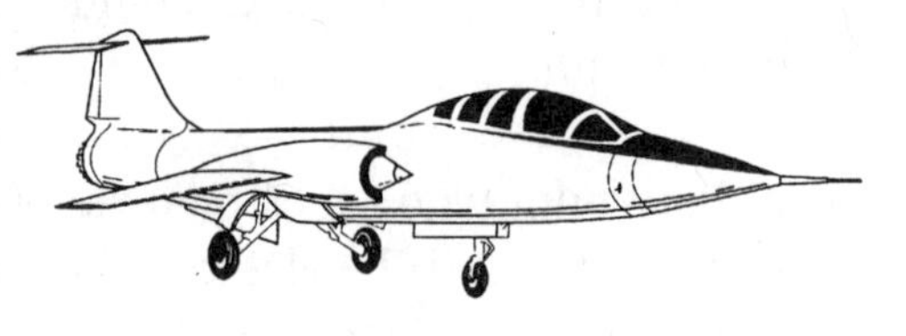

If a crash is inevitable, try to (1) hit the softest object in the vicinity, (2) as slowly and gently as possible.

When violating some silly rule, make your performance flawless (for example, if you fly under a bridge, try not to hit the bridge).

When flying VFR, stay out of the fluffy little clouds, for mountains sometimes lurk inside them.

When the tanks are half empty, it's past time to review your plan.

Plan ahead! Keep checking! If you find yourself on the ground or sitting in your rubber raft – looking up in the sky where your aircraft was a few minutes ago – it's too late to check your fuel gauge.

Don't crash while trying to fly the radio. Military aircraft fly because of the principle discovered by Bernoulli, not Marconi.

Remember, in combat your aircraft is not a tank. Your windshield isn't hi-tech plastic that bullets bounce off of.

Never let your aircraft take you somewhere your brain didn't get to five minutes earlier. Remember that you fly your aircraft with your *brain*, not with your hands.

In a military aircraft in flight, if something is (1) red, (2) yellow, or (3) dusty, never touch it without a lot of forethought.

If you value longevity, watch your six.

Remember, in combat your aircraft is not a tank. Your windshield isn't hi-tech plastic that bullets bounce off of.

Aircraft weight/temperature/altitude charts are *tools*, not *rules*. Play it safe and make two trips.

Before takeoff always pause and ponder: "How much does all that Grunt stuff in the cabin *really* weigh?"

Before takeoff always pause and ponder: "How much does all that Grunt stuff in the cabin *really* weigh?"

The two crucial absolutes in Military Aviation: (1) in airplanes, keep your airspeed up, and (2) in helicopters, keep your rotor RPM up. Otherwise, the Earth will rise up and smite thee.

Never forget the six most useless things in Military Aviation:

1. The approach plates you forgot to bring.
2. The fuel you've burned.
3. The airspeed you had.
4. The altitude above you.
5. The runway behind you.
6. A tenth of a second ago.

When flying, pilots should try to stay in the middle of the air. The edges of the air can be recognized by the appearance of trees, towers, buildings, telephone poles, wires, the ground, the sea, and mountains. It's very hard to fly beyond the edges of the air.

Fighter and Attack pilots: when your fear of the aircraft exceeds your fear of the ejection seat, it's time to say goodbye.

Airspeed, Altitude, Brains. You need at least *two* at all times.

An E-3 Sentry returns to Earth (photo courtesy of U.S. Air Force).

– Murphy's Aviation Ironies –

In an emergency when you've run out of bright ideas, luck may be an acceptable substitute. But, in the long run, trusting luck alone is not conducive to longevity.

In military aviation you start with (1) a *full bag* of luck, and (2) an *empty bag* of experience. The trick is to fill your bag of experience before you empty your bag of luck.

Combat flying isn't like a video game. When flying in combat, you can't simply push a button and start over.

When a warrior flies in combat, (1) being *good* and (2) being *lucky* often isn't quite enough.

Day or night, the most crucial radio frequencies will be illegible.

Combat *flying* isn't dangerous – *crashing* is what's dangerous.

A thunderstorm usually isn't quite as bad on the inside as it looks on the outside – it's worse.

There are three simple easy-to-understand rules for making smooth landings – but nobody knows what they are.

Everything that goes up must come down. Going up is usually easy. It's the manner of *coming down* that can spoil your day.

In a crisis it's always better to (1) break ground and head into the wind, than to (2) break wind and head into the ground.

In a crisis it's better to (1) be on the ground, wishing you were flying, than to (2) be flying, wishing you were on the ground.

The farther you fly over the mountains IFR at night, the stronger the strange fuselage vibrations will become.

A thunderstorm usually isn't quite as bad on the inside as it looks on the outside – it's worse.

Combat flight experience is something you never have enough of until *after* you desperately need it.

In retractable-gear aircraft, if it takes over 80% power to taxi, you likely have landed gear-up.

There are three simple easy-to-understand rules for making smooth landings – but nobody knows what they are.

Experience shows that a smooth landing is *mostly luck*, two in a row is *all luck*, and three in a row is *lying*.

When flying, you're never lost if you don't care where you are.

In combat it's true that more aircraft are downed by a shortage of spare parts than by enemy fire. The difference is that pilots and aircrewmen rarely *die* due to a shortage of spare parts.

A new CV-22 Osprey tilt-rotor machine can (1) hover like a helicopter and also (2) fly like a fixed-wing airplane (photo courtesy of U.S. Air Force).

A competent military pilot has mastered a host of complex skills. A competent aircrewman can perform myriad functions. Yet, none of these skills and functions guarantee survival in combat.

– Murphy's Aviation Predictions –

Heavier-than-air flying machines are impossible.
[Lord Kelvin, President of the Royal Society, writing in 1895]

It is complete and utter nonsense to believe that flying machines will ever work.
[Sir Stanley Mosley, philosopher, writing in 1905]

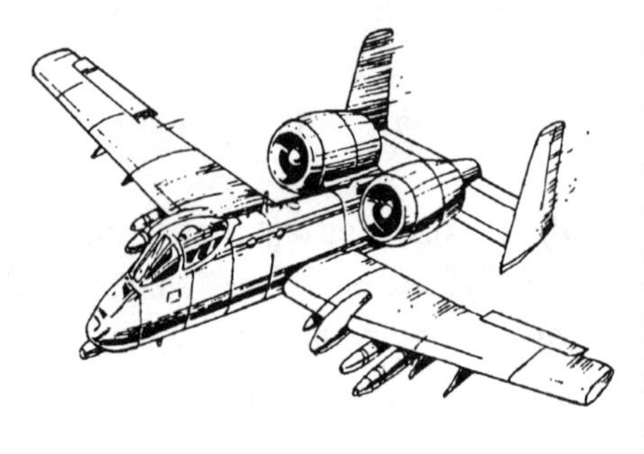

The helicopter had no future, so we dropped it. The helicopter does, with great labor, only what the balloon does without labor. The helicopter is no more fitted than the balloon for horizontal flight. If its engine stops, it must fall with deathly violence, for it can neither glide like an aeroplane nor float like a balloon. The helicopter is easier to design than the aeroplane, but it is utterly worthless.

[Wilbur Wright, co-developer and pilot of the world's first successful heavier-than-air, powered, controllable airplane in 1903, writing in 1909]

We do not consider that aeroplanes will be of any possible use for war purposes.

[Report of the British Secretary of State for War, 1910]

The airplane is an invention of the devil.

Airplanes are interesting toys, but they are of no military value.

[Gen. Ferdinand Foch, French military strategist, 1911]

The aeroplane is an invention of the devil. It will never play any part in the defense of the nation, my boy!

[Sir Samuel Hughes, Canadian Minister of Defense, 1914]

Aaaaahhh, sh--! *[an expletive]*

[According to McDermott Associates (specialists in cockpit voice recorders) the most common "final words" on cockpit voice recorders from aircraft that have crashed – voiced without panic or fear; merely the pilot's admission that he (1) knew he was going to crash, (2) had done everything humanly possible in an effort to prevent it, and (3) was resigned to his fate]

Murphy's Special Laws for Helicopters

The previous chapter, "Murphy's Laws of Combat for Aviation," applies to both fixed-wing aircraft and helicopters. Yet, helicopters are *different*. They kill you *quickly*. No matter how skilled an aerial warrior may be, no matter how lucky he is, no matter how much his mother loves him, helicopters can kill him in the twinkling of an eye. On the *ABC Evening News* on 16 February 1971, anchorman Harry Reasoner explained to a nationwide audience:

> A helicopter does not want to fly. It is maintained in the air by a variety of forces and controls working in opposition to each other. And if there is any disturbance in this delicate balance, the helicopter stops flying immediately and disastrously.
>
> There is no such thing as a gliding helicopter! That is why being a helicopter pilot is so different from being an airplane pilot. Airplane pilots are open clear-eyed buoyant extroverts, and helicopter pilots are always brooding introspective anticipators of trouble. Helicopter pilots know that if something bad has not yet happened – it is about to.

A helicopter can kill you more quickly than any other instrument ever conceived by man. And in combat, helicopters fly to evil places where bad guys can kill you quickly too. Therefore, Murphy warns all helicopter pilots and aircrewmen to remember five eternal truths about their flimsy fling-wing contraptions:

1. In a helicopter, eternal vigilance is the price of survival.
2. There's no such thing as a gliding helicopter.
3. The safest helicopter is the one that can barely kill you.
4. You're always a student in a helicopter.
5. If at first you don't succeed, ***never*** try autorotations again!

Helicopters don't last very long. This chunk of twisted aluminum was once a big CH-46 Sea Knight helicopter (photo by the author, Marion Sturkey).

To give helicopter crews a glimmer of hope of surviving, Murphy offers invaluable insight about their accident-prone flying machines. Any warrior stupid enough to consider flying, or riding, in a helicopter should carefully study the following sub-chapters:

1. Murphy's Helicopter Philosophy
2. Murphy's Admonitions for Helicopter Crews
3. Murphy's Simple Rules for Learning to Fly Helicopters

– Murphy's Helicopter Philosophy –

Any mechanical contraption that attempts to *screw* its way into the sky is doomed to failure.

According to laws of physics and aerodynamics, helicopters can't actually fly. They're simply so ugly that the Earth repels them.

In helicopters, the foreseeable future is the next ten seconds. Long range planning is the next two minutes.

> In helicopters, the foreseeable future is the next ten seconds.

Helicopters are tricky machines. Pilots and aircrewmen measure their lives in *days* instead of *years*.

When helicopter "wings" are (1) leading, (2) lagging, (3) precessing, and (4) flapping, sinister forces are at play.

The terms *protective armor* and *helicopter* are mutually exclusive.

The phenomena known as (1) Vortex Ring State, (2) Retreating Blade Stall, and (3) Power Settling are nothing more than fancy ways to describe *instant death*.

Helicopter crews fly with an intensity akin to "spring loaded" while waiting for pieces of their craft to fly off.

You can always identify a helicopter crewman in a car, or a boat, or a train. He (1) never smiles, he (2) listens to the machine, and he (3) always hears something he thinks is not quite right.

> There are two types of combat aircraft, (1) fighters, and (2) targets. Unfortunately, a helicopter is not a fighter.

At any small airport there are lots of old airplanes lying around, but you *never* see an old helicopter. Think about it!

Whoever said "the pen is mightier than the sword" never flew in a helicopter into a AAA and missile-threat environment.

Flying helicopters into a AAA and missile-threat environment doesn't require *courage*. It requires *stupidity*.

The terms *protective armor* and *helicopter* are mutually exclusive.

When (1) the weather is clear, (2) the rotors are in track, (3) the fuel tanks are full, and (4) all gauges are in the green, you're about to be surprised. That's what helicopters do.

There are two types of combat aircraft, (1) fighters, and (2) targets. Unfortunately, a helicopter is not a fighter.

In fixed-wing aircraft, airspeed is *life*, and (2) altitude is life *insurance*. But in helicopters, rotor RPM is ***everything!***

Sudden unexpected noises somewhere in the helicopter *always* will get the crew's undivided attention.

Sudden unexpected noises somewhere in the helicopter *always* will get the crew's undivided attention.

The three best things in life are (1) a good orgasm, (2) a good landing, and (3) a much needed bowel movement. For a helicopter crew, a successful return from an emergency night medevac is a unique opportunity to experience all three of them at the same time.

In combat, there's no such thing as a *secure* helicopter LZ. Anyone who says otherwise is selling something.

On any medevac, the amount of time you must spend on the ground in the LZ will be *directly* proportional to the intensity of enemy fire.

On emergency night medevacs, the LZ coordinates invariably will be at the junction of several maps.

In combat, there's no such thing as a *secure* helicopter LZ. Anyone who says otherwise is selling something.

Among helicopter crews, after a successful emergency ammo resupply at night, the *first* liar doesn't stand a chance.

The enormous CH-53 Sea Stallion (photo courtesy of U.S. Marine Corps).

Flying is better than riding in a vehicle, which is better than running, which is better than walking, which is better than crawling – all of which are better than an emergency night medevac in a helicopter, although it's *technically* a primitive form of flying.

Will Rogers never met a fighter pilot.

Fighter pilots make *movies*, but helicopter crews make *history*.

– Murphy's Admonitions – for Helicopter Crews

For military pilots and aircrewman, helicopter time in your logbook is akin to S.T.D. in your Health Record.

Flying in a helicopter is about the same as masturbating – it may be fun at the time, but it's nothing to brag about in public.

Rotor RPM *must* be kept within the green arc! Failure to heed this admonition will adversely affect the morale of the crew.

> Flying in a helicopter is about the same as masturbating – it may be fun at the time, but it's nothing to brag about in public.

When flying helicopters, the main trick is to keep the fuselage from spinning as fast as the rotors.

If everything on your helicopter is temporarily working properly, consider yourself temporarily lucky.

When flying in a military helicopter, *keep checking*. There's always something you've missed.

Running out of (1) airspeed, (2) altitude, (3) rotor RPM, (4) luck, and (5) bright ideas will spoil your day.

Running out of (1) collective, (2) pedal, (3) forward cyclic, or (4) aft cyclic are all exceptionally bad ideas.

Running out of (1) collective, (2) pedal, (3) forward cyclic, or (4) aft cyclic are all exceptionally bad ideas.

If your engines fail you have 3/10 of a second to either (1) lower the collective, or (2) begin flying like a manhole cover.

When flying helicopters, the main trick is to keep the fuselage from spinning as fast as the rotors.

In combat, helicopter crews fly on a reactive basis, so (1) eat when you can, (2) sleep when you can, and (3) defecate when you can. Your next opportunity may not come for a long, long time.

Flying a helicopter at an altitude in excess of 250 feet is considered downright foolish.

A CH-46 Sea Knight helicopter, EP-153, takes off from the USS Okinawa near Vieques, Puerto Rico (photo by the author, Marion Sturkey).

If you ditch at sea in a helicopter, get out immediately, for it will sink within 20 seconds – plus or minus about 18.

In combat, it is unwise to land your helicopter in an enemy-infested area that your fixed-wing friends have recently bombed or strafed. The bad guys won't be happy to see you.

In a military helicopter, *death* is the price a warrior pays for trying to look Sierra Hotel.

– Murphy's Simple Rules – for Learning to Fly Helicopters

If a warrior can (1) drive a car or (2) ride a bicycle, there's no reason why he shouldn't give helicopters a try. Murphy has compiled eight simple and easy-to-understand rules for red-blooded warriors who are non-practitioners in the world of helicopters.

A warrior who wants to fly helicopters may be an airplane pilot or aircrewman with an urge to expand his skills. Or, he may be a Grunt, a professional population control specialist, who dreams of zipping around in a helicopter during wars and firefights. On the other hand, a warrior may think that newfound fame – *or infamy* – as a helicopter pilot will enhance his success rate when he tries to coerce, corrupt, and seduce members of the opposite sex.

> If a warrior can (1) drive a car or (2) ride a bicycle, there's no reason why he shouldn't give helicopters a try.

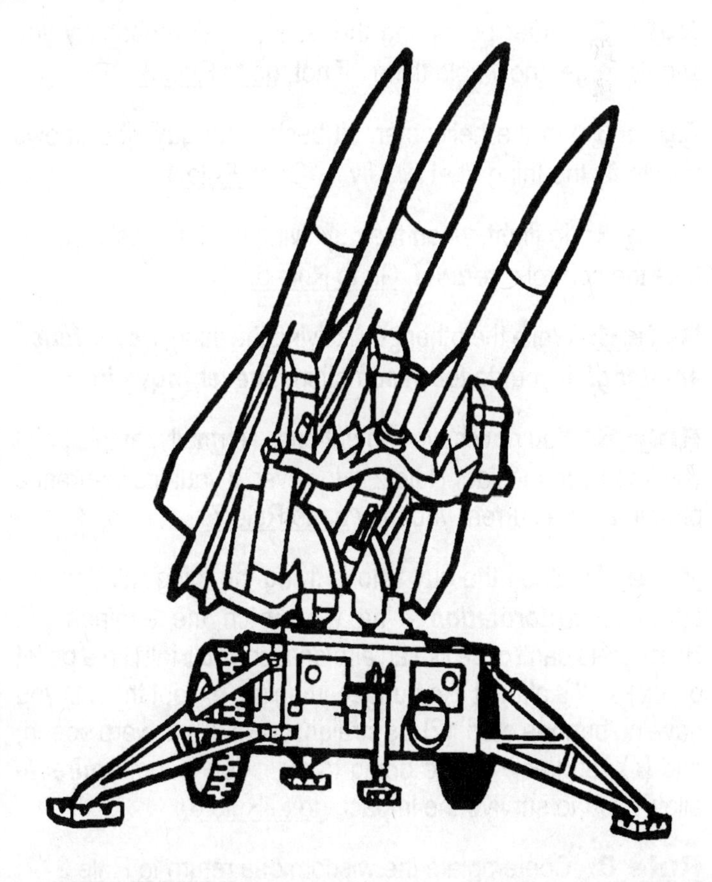

If a military helicopter is buzzing around over enemy territory, and the crew spots a surface-to-air-missile battery like this on the ground below them, it guarantees an instant cure for constipation.

There are many valid reasons why a warrior might yearn to learn how to fly helicopters. For these slow-learners, ***Murphy's Simple Rules for Learning to Fly Helicopters*** are displayed below:

– Murphy's Simple Rules –
for
Learning to Fly Helicopters

Rule 1: ***Think about it!*** Physicists, aerodynamic experts, and scientists have no idea what holds a helicopter up. But, whatever it is, it could stop at any time. Go to Rule 2.

Rule 2: After pondering the issue, in all probability you should forget the whole thing. If not, go to Rule 3.

Rule 3: In the helicopter, sit beside the guy who knows how to fly the thing. Let *him* fly it. Go to Rule 4.

Rule 4: In flight, when the guy flying the thing asks you to take the controls, ***refuse!*** Go to Rule 5.

Rule 5: While the other guy is flying the thing, ***never touch anything!*** If you do touch something, ***never move it!***

Rule 6: You now have two options. Normally, at this point you should reconsider Rule 2. However, if your life insurance premiums are current you may go to Rule 7.

Rule 7: Soon the guy who is flying the thing will demonstrate an ***autorotation*** – he will switch the engines off. Helicopters can't glide like an airplane, so you'll fall like a pallet of bricks. It's almost like bungee jumping, except that (1) you have no bungee cord, (2) it's straight down, (3) at warp speed, and (4) you know you're going to die. Yet, if some ***miracle*** allows you to survive the impact, go to Rule 8.

Rule 8: Contemplate the wisdom of a return to Rule 2.

Murphy's Military Definitions for Aviation

Heads up, aviation warriors! If *staying alive* in aerial combat is of concern to you, this is crucial scientific stuff. Even in peacetime, military aircraft have a penchant for perverse and evil habits. They come unglued in flight. They run out of gas. They butt heads with mountains at night. Such idiosyncrasies adversely affect the longevity of those who ride or fly in flimsy aerial contraptions. And, don't even *mention* helicopters! If Grunts knew all that Murphy knows about military helicopters, they'd never ride in one again.

> Even in peacetime, military aircraft have a penchant for perverse and evil habits.

In this chapter, Murphy defines the things that keep aerial warriors alive – or kill them – such as *Hydroplane* and *Dead Reckoning*. *Pucker Factor* is scientifically explained. If a warrior is confused about the *Bang-Stare-Red Theory* or the mystery of *Retreating Blade Stall*, Murphy has the right answers.

Also, many Magnificent Grunts may desire insight into military aviation, the most perilous mode of transportation on Earth. If so, they should delve into this chapter. Thereafter they may conclude that early retirement isn't a bad idea. ***Murphy's Military Definitions for Aviation*** are displayed below:

A-Model: (1) An underpowered experimental aircraft prototype. (2) A primitive contraption a warrior should *never* ride or fly in.

Accident Investigation Board: ***Six men***, who take ***six months***, to debate what the deceased aircrew did during the last ***six seconds*** of their lives – in the rain, at night, in the mountains, under fire.

Acey-Deucy: A competitive endeavor designed to weed out any aerial warriors who lack the killer instinct necessary for combat.

ADF: An aircraft navigation instrument of last resort.

Aeronautics: (1) Neither an industry nor a science. (2) A miracle.

Air Medal: A military award bestowed upon pilots and aircrewmen who blunder into a perilous situation in combat and – by virtue of dumb luck alone – manage to survive.

– Common Military Fixed-Wing Aircraft –

AV-8B Harrier: A really cool-looking jet VTOL aircraft with *illusions* of being a (1) helicopter, and maybe even a (2) fighter.

A-10 Thunderbolt: A wicked and mean low-flying warthog equipped with (1) a titanium bathtub, and (2) lots and lots of ammo.

B-52H Stratofortress: (1) A huge 50-plus-year-old strategic bomber in which – after four engines have failed – the crew still has four more engines that have not. (2) Proof of the hypothesis that, if you glued enough engines onto it, even a brick could fly.

F-15A/E Eagle: A sleek aircraft whose most useful function is to astound air-show attendees with a nifty vertical vanishing act.

C-130A/J Hercules: A transport aircraft (1) designed in the 1940s, (2) in service in the 1950s, and (3) still being *built* a half-century later.

F-16C Fighting Falcon: (1) A concession to economics. (2) The first "modern" United States combat aircraft with less than two engines.

F-117A Nighthawk: The plane with the smallest, and the lightest, ordnance load in the entire U.S. Armed Forces.

F/A-18 Hornet: A sleek supersonic flying machine primarily useful for enemy surface-to-air-missile target practice.

F-22A Raptor: A mean and stealthy super-cruising aircraft designed to assist enemy aerial warriors who wish to die for their country.

F-35 Lightning: The expensive "joint strike fighter" so woefully anemic that three variants are needed to make it militarily viable.

A troop-carrying CH-46E Sea Knight helicopter from HMM-262 launches from the flight deck of the USS Essex (photo courtesy of U.S. Navy).

Airspeed: Expressed in knots, the speed of a military aircraft relative to the air mass through which it travels (deduct 40% when listening to civilian pilots or REMFs).

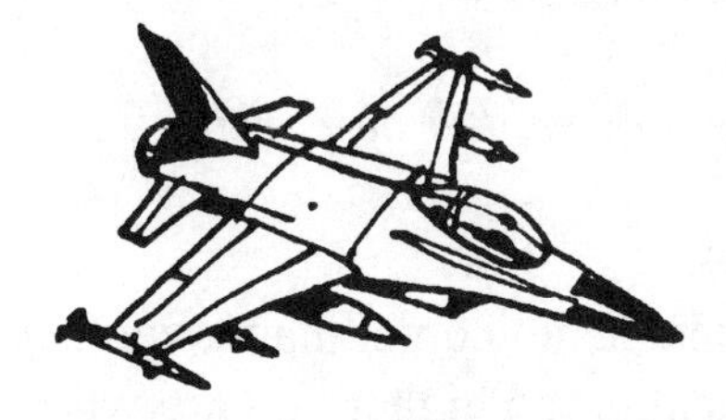

All-Weather Close Air Support: Superior and overwhelming aerial fire support for Grunts, available 24 hours per day *except* (1) during inclement weather, and (2) at night.

Alpha-Mike-Foxtrot: A common verbal farewell to (1) any despicable person, or to (2) an enemy aircraft you have shot down.

Alternate Airport: A carefully selected airport located at least 50 miles beyond the maximum range of an IFR aircraft.

Alternate Airport: A carefully selected airport located at least 50 miles beyond the maximum range of an IFR aircraft.

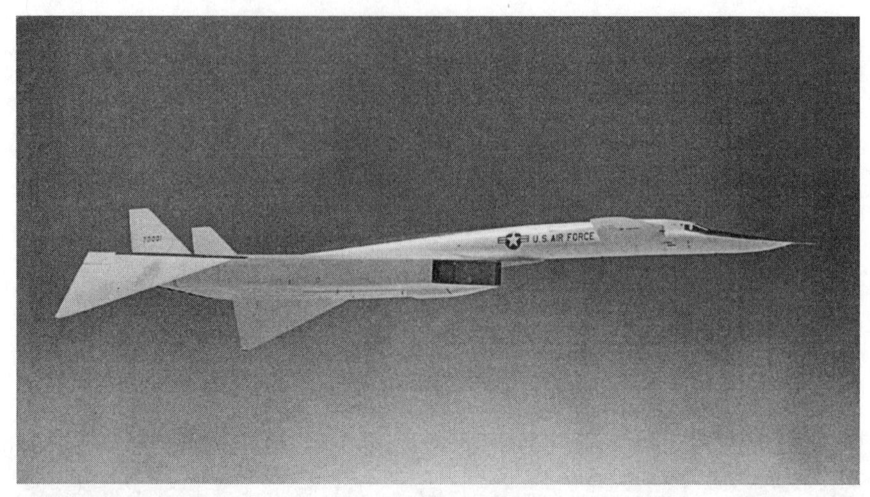

The late and great XB-70 Valkyrie, a compression-lift Mach 3 bomber, one of the most beautiful aircraft ever built (photo courtesy of U.S. DOD Visual Information Center).

Altimeter: A barometric pressure sensing device that indicates height above *sea level* – and therefore *useless over land.*

> Radar Altimeter: (1) Better than the regular altimeter. (2) An electronic sensing device that indicates height above either the land or the sea. (3) An instrument that works just fine – unless you're flying IFR toward the side of a sheer cliff.

Attitude Indicator: A cockpit instrument that *better be working* while a warrior is flying IFR, or at night.

Autorotation: A terrifying and inexplicable helicopter maneuver that you only get *one chance* to try. Ironically, the higher you are when the terror begins, the better. (also, see "Pucker Factor")

Back Side of the Power Curve: (1) An evil regime of *slow flight* defined by obscure laws of aerodynamics that nobody understands. (2) The side of the curve you don't want to be on.

> Barrel Roll: The best way to move large beer containers.

Bang-Stare-Red Theory: A scientific truth which substantiates that (1) the louder a sudden *bang* in an aircraft, the quicker the pilot's eyes will be drawn to the gauges, and (2) the longer the pilot stares at the gauges, the quicker the needles will move from the green arcs into the red arcs.

Bank: The generic name for the civilian institution that holds the lien on POVs driven by military pilots and aircrewmen.

Barrel Roll: The best way to move large beer containers.

Cloud: A beautiful opaque meteorological phenomenon within which mountains frequently lurk.

Carburetor Icing: In piston-powered manned aircraft or UAVs, a phenomenon which occurs when the fuel tanks mysteriously become full of air. (also, see "Engine Failure")

CAVU: (1) The stuff that aeronautical dreams are made of. (2) What Heaven surely must be like.

Chicken Plate: (1) Part of the standard combat attire for helicopter crews. (2) Something that can't be ordered in a restaurant.

Chip Detector Light: A *really* evil thing when you're flying IFR.

Cloud: A beautiful opaque meteorological phenomenon within which mountains frequently lurk.

Cluster Bombing: An aerial bombardment technique that is 100% accurate, because the bombs *always* hit the ground in clusters.

Combat, Aerial: What an aerial warrior is in when *orange baseballs* are zipping past, and sometimes through, his aircraft.

Copilot: A useless cockpit occupant *until* he spots closing traffic at 12 o'clock (after which he's an ignoramus for not seeing it sooner).

Crab: The squadron Safety Officer.

– Common Military Helicopters –

AH-1Z Super Cobra: (1) A *skid-mounted* flying machine unable to adapt to *wheel* technology. (2) A mini-attack helicopter that added more rotor blades in an attempt to look like an Apache.

AH-6D Apache Longbow: The all-time least favorite helicopter of all bad guys who are crewmen in tanks and armored vehicles.

CH-46E Sea Knight: An underpowered, unreliable, and antiquated flying contraption known by its survivors as the "frog" or "shuddering [expletive]-house." Just say ***Hail Mary*** and climb aboard.

CH-47D Chinook: The big, bold, and bad combat helicopter that its younger brother, the CH-46E, should have been.

CH-53E Sea Stallion: Proof of the theoretical aeronautical concept that, if you keep on adding (1) more and more engines, and (2) more and more rotor blades, *anything* will eventually fly.

UH-1N Huey: A primeval and tiny heavier-than-air rotary-wing flying machine apparently designed before the invention of *wheels*.

UH-60L Blackhawk: The same as a Huey on wheels and steroids.

MV-22 Osprey: (1) An expensive, futuristic, experimental, hi-tech bird of ***prey***, in which you should ***pray***. (2) Often a helicopter, often an airplane, but sometimes just a pile of burned rubble.

Crash: Nature's way of reminding Sierra-Hotel military pilots to watch their airspeed.

Cruise Box: A gigantic heavy footlocker, loaded aboard a military aircraft, which reduces the passenger load by at least two Grunts.

Dead Reckoning: (1) You reckon correctly, or you are. (2) The least accurate method of aerial navigation.

Emergency Extraction: (1) What helicopter aircrews often do in combat, usually at night. (2) A guaranteed remedy for "tired blood."

Crash: Nature's way of reminding military pilots to watch their airspeed.

Emergency Medevac (by helicopter): (1) A priority flight to save the life of a wounded brother-in-arms. (2) A quick cure for constipation.

Engine Failure: An aerial phenomenon which occurs when all fuel tanks somehow become full of air.

Experienced Crew: A combat aircrew that has survived long enough to recognize their mistakes when they make them again.

FAA Motto: We're not happy until you're not happy.

Famous Last Words, Most: "Don't worry about the weight, it'll fly."

Firewall: A metallic structure, behind a forward-mounted engine, designed to channel fire and smoke into the cockpit.

Experienced Crew: A combat aircrew that has survived long enough to recognize their mistakes when they make them again.

Flying, Day: (1) A thing hard to do without feathers. (2) The common term for the illusion of immortality. (3) The ability to throw yourself through the sky and avoid hitting the ground.

Flying, Night: The same thing as *Flying, Day* – except that you can't see where you're going.

FOD-Burger: An inedible substance, the ingestion of which hopefully will be discovered *prior* to attempted flight.

A Chinese fighter, the Jian F-8 II Apollo, known by NATO as a Finback-B (photo courtesy of U.S. DOD Visual Information Center).

Fuel: A limited resource without which the crew and passengers in an aircraft (1) become pedestrians, or (2) become deceased.

Gravity: The primary cause of most military aircraft crashes.

G-Suit: Clothing designed to prevent untimely aerial napping.

Glide Distance: Half the distance from any aircraft in distress to the nearest suitable landing area.

Glider: An airborne military aircraft, the fuel tanks of which have become full of air.

G-Suit: Clothing designed to prevent untimely aerial napping.

GPS: (1) The aviation acronym for "Going Perfectly Straight." (2) Also, the common name for the electronic black-box gizmo that lets the pilots go perfectly straight.

Gravity: The primary cause of most military aircraft crashes. Gravity may not be fair, but (1) it's the law, (2) it's not subject to repeal, and (3) it's forever.

Headwind: (1) The guaranteed result of any attempt to stretch fuel. (2) A meteorological occurrence on most long over-water flights.

Helicopter: The generic name for a heavier-than-air, vertical takeoff, flying machine comprised of thousands of parts, all of which rapidly spin in opposite directions, constantly striving to tear themselves apart, and often succeeding.

HIGE: The best place for a helicopter crew to hover.

HOGE: An invisible and mysterious aerodynamic capability that helicopter crews always want to *have*, but never want to *use*.

Glider: An airborne military aircraft, the fuel tanks of which have become full of air.

Hover: A type of flying practiced by helicopter crews that have no specific place to go.

Hydroplane: A flying machine designed to land on wet runways.

IFR: (1) Not nearly as good as regular VFR. (2) In common usage, the official acronym for "I follow railroads." (3) A tricky method of flying by needle and horoscope.

Instrument Flying: (1) An unnatural act. (2) Not a very good idea. (3) How a warrior flies when he can't fly like he wants to fly.

Hydroplane: A flying machine designed to land on wet runways.

Jet Engine: An expensive tool for converting kerosene into noise.

Landing Gear Handle: The cockpit handle that a wise military pilot will lower to the *down* position immediately after his inadvertent gear-up landing.

MARCAD (or NAVCAD): An extinct USMC/USN flight training program that produced (1) *twice* the pilot, at (2) *half* the price.

Mayday! Mayday! Mayday!: A verbal notice that prayer – while it may not help – is still an excellent idea.

Meatball: For military pilots, a longevity-related phenomenon to be closely watched, not eaten.

Mile High Club: (1) If you don't already know, you aren't already in it. (2) A club that requires co-conspirators of the opposite sex.

Minimums: After crashing and surviving, the altitude below which the pilot must swear he did not descend while flying IFR.

– Murphy's Landing Analysis –

Landing: A technique for falling out of the sky with style.

Good Landing: Any landing after which (1) the aircraft doors will still open, and (2) the aircraft can be salvaged.

Great Landing: Any landing after which (1) the aircraft doors will still open, (2) the aircraft can be salvaged, and (3) all pilots and passengers can walk away without assistance.

Mixture, Lean: Beer with very little alcohol content.

Mixture, Rich: High-alcohol-content beer that red-blooded warriors order at the *other guy's* promotion party.

Mile High Club: If you don't already know, you aren't already in it.

Nanosecond: The time delay built into stall warning systems.

Navigation, Aerial: The scientific process used by a combat aircrew to get from Point A to Point B – while *trying* to get to Point C.

OBE: (1) An untenable condition that occurs when a military aircraft travels faster than the brain of its pilot. (2) Task overload. (3) A condition closely akin to the snakes-in-the-cockpit syndrome.

Minimums: After crashing and surviving, the altitude below which the pilot must swear he did not descend while flying IFR.

Pilot: A confused person who (1) talks about *women* when he's flying, and (2) talks about *flying* when he's with women.

Precision Bombing: Hi-tech and smart-weapon aerial bombardment accurate to within plus or minus seven miles – more or less.

Preflight Planning: (1) A time consuming exercise in futility. (2) A trite endeavor to which Sierra-Hotel military pilots give lip-service only when the Safety Officer is on the prowl.

– Murphy's "Pucker Factor" Formula –

Pucker Factor: (1) The formula **(T x I x R over H)** which determines the contraction force of an aviator's *Gluteus Maximus muscles*. (2) The common term for the degree of contraction of these muscles in times of peril. (3) In technical aviation language, the mathematical formula which determines the amount of seat cushion that will be sucked up into the butts of pilots and aircrewmen when they are under enemy fire. (4) On a 1-to-500 ascending scale, ***Pucker Factor*** may be scientifically and mathematically quantified as follows:

T (number of tracers headed your way),
multiplied by I (your interest in staying alive),
multiplied by **R** (your rate of descent),
divided by **H** (your height above ground),
equals your ***Pucker Factor***

Range: (1) In *theory*, the distance an aircraft can fly. (2) In *reality*, the erroneously-stated distance 50 miles beyond the point where all fuel tanks will become totally full of air.

An AH-1W Super Cobra – called a "Snake" – from HMM-162 prowls the air over the Atlantic Ocean (photo courtesy of U.S. Marine Corps).

SAR: A type of mission you hope you're never the objective of.

Retreating Blade Stall: (1) The aerodynamic nemesis of helicopters that fly too fast before crashing. (2) Something *really* repulsive. (3) The primary cause of insomnia among helicopter crewmen.

Roger: The *standard* radio response from military pilots when they don't know the *proper* radio response.

Roll: A design priority for all transport helicopters.

Running Takeoff: A nifty practice maneuver, but if a helicopter *has* to do it, you don't want to be riding in it.

SAR: A type of mission you hope you're never the objective of.

Stall: A warrior's best technique to thwart proposals of matrimony.

Separation: The condition achieved when two or more aircraft fail to collide in flight.

Sierra-Hotel: (1) What an aerial warrior says in correspondence, or in mixed company, when he can't say what he *really* wants to say. (2) A military warrior's best *politically incorrect* synonym for: Outstanding! Aggressive! Supremely skilled!

Single Engine Capability: A "level flight" capability which most multiengine aircraft have – until they try to land.

Slip: Flimsy civilian undergarments worn by some women.

Spoilers: Members of the Accident Investigation Board.

Stall: A warrior's best technique to thwart proposals of matrimony.

Tail Rotor: (1) The delicate helicopter rotor that is magnetically drawn toward trees, stumps, poles, wires, and other obstructions to flight. (2) The fragile little rotor that – unlike the big main rotor that can chop down hickory trees – will self-destruct if it hits anything bigger than a honeybee.

Spoilers: Members of the Accident Investigation Board.

Tailwind: The result of eating far too many beans and other leguminous food products.

Terminal Forecast: A complex horoscope with lots of numbers.

Terminal Forecast: A complex horoscope with lots of numbers.

Thunderstorm: (1) Mother Nature's way of saying, "Up yours!" (2) The common term for *cumulo-securus* clouds.

Weather: Next to gravity, the biggest cause of aircraft crashes.

Translational Lift: (1) For helicopter crews, a *very good thing.* (2) A phenomenon attributed to black magic. (3) An aerodynamic wonderland that vanishes, unfortunately, when a pilot tries to land.

Turn and Slip Indicator: A cockpit instrument of no use to pilots.

Useful Load: The total volumetric capacity of any military aircraft, regardless of the total weight.

VFR: (1) The meteorological conditions under which members of an aircrew can see what they collided with. (2) The simple rules a pilot can "declare" if a proposed IFR procedure is too complex.

Special VFR: (1) Even better that regular VFR. (2) The rules a pilot can "declare" when nothing else will work.

VFR: The simple rules a pilot can "declare" if a proposed IFR procedure is too complex.

Weather: Next to gravity, the biggest cause of aircraft crashes.

Whiskey-Tango-Foxtrot?: (1) An aerial warrior's *polite* radio or ICS query. (2) An in-flight question most appropriate when the crew of another aircraft is engaged in exceptionally stupid conduct.

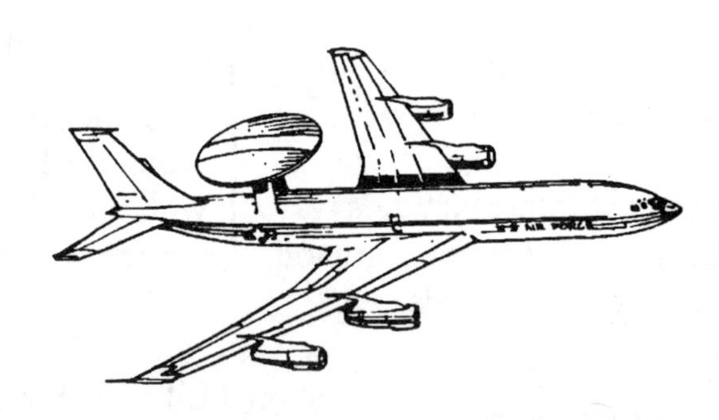

Murphy's Oaths of Enlistment for the U.S. Armed Forces

Many years ago before Murphy enlightened the world with his wisdom, professional warriors always swore allegiance to something or somebody. Most often they pledged eternal loyalty to their military leader: Genghis Kahn, Alexander the Great, or whomever. Years rolled by, and in 1775 the British colonies in North America prepared to wage war for independence from England. The new Continental Army established a formal ***Oath of Enlistment*** for recruits:

> I, ____________________, have this day voluntarily enlisted myself, as a soldier, in the American Continental Army, for one year unless sooner discharged; and I do bind myself to conform, in all instances, to such rules and regulations as are, or shall be, established for the government of the said Army.

Today, over 200 years later, those who enlist in the U.S. Armed Forces must take an oath to support and defend a *document*, the Constitution of the United States of America:

> I, ____________________, do solemnly swear, or affirm, that I will support and defend the Constitution of the United States against all enemies, foreign and domestic; that I will bear true faith and allegiance to the same; and that I will obey the orders of the President of the United States and the officers appointed over me, according to regulations and the Uniform Code of Military Justice, so help me God.

Warriors in the Air Force National Guard or the Army National Guard must pledge allegiance to the United States, of course. However, they also owe allegiance to their state and its governor, so they must take the following oath. It incorporates the name of the state in which they are enlisting:

> I, ________________, do solemnly swear, or affirm, that I will support and defend the Constitution of the United States and the State of ________________ against all enemies, foreign and domestic; that I will bear true faith and allegiance to the same; and that I will obey the orders of the President of the United States and the Governor of ________________ and the orders of officers appointed over me, according to law and regulations, so help me God.

Unlike the Armed Forces of many other countries, the United States has an oath for those who enlist (see the previous page), and a separate oath for commissioned officers. Those being commissioned must take the following oath, which is mandated by Section 3331, Title 5, United States Code:

> I, ________________________, do solemnly swear, or affirm, that I will support and defend the Constitution of the United States against all enemies, foreign and domestic; that I will bear true faith and allegiance to the same; that I take this obligation freely, without any mental reservation or purpose of evasion; and that I will well and faithfully discharge the duties of the office on which I am about to enter, so help me God.

However, ***Murphy sees a big, big problem*** with these present-day oaths. He knows the one-oath-fits-all mentality won't work. Because of vast evolutionary changes, Army doggies and Navy squids have nothing in common. Plus, Air Force zoomies and Marine Corps jarheads exist on the extreme opposite ends of the military cultural

spectrum. Each branch of the U.S. Armed Forces has its own mission. Therefore, each should have its own oath.

Fortunately, Murphy has solved this thorny dilemma. He has created four new and unique ***Oaths of Enlistment***, one for each branch of the U.S. Armed Forces:

Uncle Sam wants **YOU**!

– United States Navy –
Murphy's Oath of Enlistment

I, ________________________ (name of prospective Swabjockey), instead of completing the remainder of my prison sentence, agree to participate in four (4) years of alternative service in the United States Navy because (check one or more):

___ The Air Force is too intellectual for me.

___ I want to hang-out with Marines without having to *be* one.

___ I enjoy all water sports.

___ Other (explain): ____________________________

I fully understand (1) that for four years I must wear clothing that went out of style in the 1960s; (2) that my name will be stenciled on the butt of every pair of trousers I own; (3) that for some reason I must learn and speak a different language than the rest of the English-speaking world; (4) that I may take pride in the fact that all Navy acronyms, rank, insignia, and everything else are different from those of the *real* military services; (5) that when off base I will be mistaken for the Good Humor man in summer and the Waffen SS in winter; (6) that I must hone my coffee-cup-handling skills to the point were, aboard a destroyer in a typhoon with 100 foot seas, I will not spill a drop; (7) that I will muster *(whatever that is)* at 1100 each workday unless I am buddy-buddy with the Chief, in which case I may sleep until afternoon T.V. game-show time; (8) that if I display initiative I shall be promoted to Nautical Paint-Picker; and (9) that if I fail to display initiative I shall be cast into the pit with all other dark and slimy squid-like creatures, and we shall be banished to the murky depths of the ocean where we will be unable to mate and cause further damage to the world's gene pool, and where normal human beings will not have to associate with us.

____________________________________ (signature of parolee)

– United States Army –
Murphy's Oath of Enlistment

I, ______________________ (name of Rambo wannabe), agree to forfeit four (4) years of my pitiful and worthless life to the United States Army because (check one or more):

___ My ASVAB score was far too low for the Air Force.

___ The Marines may be OK – but I'm not that stupid.

___ Yachting with the Navy sounds like lots of fun, but I can't swim.

___ Other (explain):___________________________________

I fully understand (1) that after completion of boot camp and basic sexual sensitivity training I will be entitled to attend a different Army school each month, notwithstanding my lack of academic ability and intellect; (2) that when not in school my "work" hours will be 1000 to 1500 daily, Monday through Thursday, with time off for brunch as necessary; (3) that I must make a good faith effort to induce my wife or girlfriend – if any – to stay at home so that she will not leave me for a smarter Air Force guy or a better looking Sailor; (4) that for four years I shall daily claim that I am a mean and lean and green killing machine, because my drill sergeant will insist that I do so; (5) that, if I have less than three General Court Martials, I will be promoted to Bureaucrat E-7 within two years; (6) that the only "action" I am likely to see will be a verbal reprimand for sexual harassment; (7) that I must strive to maintain an authoritative appearance, even though I am required to accomplish nothing; and (8) that I shall be entitled to wear on my uniform a vast assortment of emblems, crests, badges, patches, and shiny dangling doo-dads in quantities that only a Dollar Store owner can appreciate.

____________________________ (signature of delusional idiot)

– United States Air Force –
Murphy's Oath of Enlistment

I, _______________________ (name of potential Zoomie), being incapable of finding a suitable job in the private sector, eagerly look forward to four (4) years of absolute leisure in the United States Air Force because (check one or more):

___ I am much too smart for the Army.

___ I am much too delicate for the Marines.

___ The Navy is too "corporate" for me.

___ Other (explain):___________________________________

I fully understand (1) that after completing basic training *(snicker)* I will be known as a mean, lean, donut-eating chairborne ranger; (2) that I may address my subordinates, peers, and superiors by their first names because we are not *really* in a military service; (3) that I may sit behind a desk all day, every day, and take credit for all work – if any – done by others; (4) that I should make a good faith effort to clean my knife before stabbing the next person in the back; (5) that I may remain flaccid and avoid all forms of demeaning physical exercise and/or exertion, exclusive of the annual two minute bike-riding PT test; (6) that I will never be exposed to any form of danger or threat of violence or harm; (7) that if I ever have to wear anything except civies, it will be a spiffy blue outfit fashioned after the uniform of a Greyhound bus driver; and (8) that my sole responsibility will be to claim that I support and defend the Constitution of the United States – even though I believe myself to be *above* all of that.

_______________________(signature of wild-blue-yonder-wonder)

– United States Marine Corps –
Murphy's Oath of Enlistment

I, ______________________ (name of Chesty wannabe), can not read or write. However, I *can* understand the English language if it is spoken very slowly. I agree to serve four (4) years of my wretched life in the United States Marine Corps because (check one or more):

___ I am intimidated by those good-looking Air Force women.

___ I like Navy boats, but I'm afraid of soap and water.

___ If MASH is reality, the Army is too formal.

___ Other (explain):______________________________

I fully understand (1) that for four years my lack of intelligence will be considered a great virtue; (2) that I must run around in circles shouting unintelligible animal noises for 14 hours each day; (3) that "**Kill, sir!**" will be an acceptable answer to any and all questions from a superior; (4) that being called a jarhead *(whatever that may be)* is a great honor; (5) that I will be stationed in the worst cesspools of the known world; (6) that I shall renounce all right to think – if capable of thinking – for the duration of my enlistment; (7) that each day I must drink, carouse, brawl, kick cats, sing obscene songs, embellish War Stories beyond recognition, and try to corrupt members of the opposite sex; (8) that I must go to any place, at any time, and destroy whatever, or whomever, I am ordered to destroy; and (9) that, if I am discharged or go AWOL prior to the expiration of my enlistment, I agree to voluntarily return to the psychiatric hospital, or certified in-patient mental health facility, from which I escaped.

__________________________ (mark or "X" of potential jarhead)

Murphy's Analysis: American Warriors and Beer

Beer and warfare always have gone hand-in-hand. The famed Roman Legions of Julius Caesar usually got wined and dined the night before battle. Beer (supplemented with wild wanton women and song) fired the spirit, deadened the senses, and led to a restful sleep. Without several kegs of beer the night before battle, Roman centurions would have had to soberly contemplate the next day's perils and hardships.

Medical science has proven that beer enhances health and vitality.

Fast-forward to today. Medical science has proven that beer enhances health and vitality. To fully understand this phenomenon, one should consider the attrition characteristics of an antelope herd. When the herd is hunted by predators it tries to run away, but the old, slow, and diseased antelopes in the rear of the herd are caught and killed. This natural attrition is beneficial. Killing the weak and diseased antelopes improves the general health of the herd. This culling process validates the "survival of the fittest" concept.

> Warriors find that drinking beer entails many societal virtues.

This same principle applies to the human brain. Science has proven that beer kills brain cells – but it kills only the *slowest and weakest* cells. With these weak brain cells eliminated our brains can function more accurately and rapidly. The more beer we drink, the more weak brain cells we eliminate. This enhances the health and efficiency of our brain.

Also, consider how the beer industry has packaged its products. There are 24 hours in a day, and 24 beers in a case. It can't possibly be a mere coincidence! Further, studies have proven that the perfect *balanced diet* is a beer in each hand. And in amorous

There are 24 hours in a day, and 24 beers in a case. It can't possibly be a mere coincidence!

matters of the heart, beauty lies in the eye of the *beer-holder*.

Warriors find that drinking beer entails many societal virtues. After a warrior reads about the so-called evils of drinking, he usually gives up reading. And if a woman drives him to drink, he usually has the decency to thank her. Plus, he learns a valuable lesson. If he always does, *sober*, what he said he would do, *drunk*, it teaches him to keep his mouth shut.

Nonetheless, a wise warrior must drink responsibly. He knows that excessive beer consumption may lead to otherwise inexplicable rug burns on his forehead. Too much beer can persuade him that he can converse with others without spitting on them. Worse yet, too much beer at night may cause a crusty old warrior to wake up in the morning next to something *really* scary, whose name and/or species he can't recall.

If a woman drives him to drink, he usually has the decency to thank her.

Murphy can't solve all these social conundrums. However, he does offer excellent advice for beer-drinking warriors:

– Symptom, Cause, and Solution Guide – for Beer-Drinking Warriors

Symptom: Your beer looks crystal clear.
Cause: The bottle is empty, friends are trying to sober you up.
Solution: Choose your friends more wisely.

Symptom: Your friends are below you, and laughing.
Cause: You are dancing on the table.
Solution: Dance in the center to avoid a fall.

Symptom: Your feet are squishy, clammy, and cold.
Cause: Beer bottle held at an improper angle.
Solution: Point open end of bottle toward ceiling.

Symptom: Your trousers are squishy, clammy, and warm.
Cause: Improper bladder control.
Solution: Be patient, in an hour or so, no one will notice.

Symptom: Your beer is tasteless, your stomach is cold and wet.
Cause: Missing your mouth with the bottle.
Solution: Go into restroom, practice in front of mirror.

Symptom: Suddenly you don't recognize anyone.
Cause: You've stumbled into the wrong bar.
Solution: Ask if anyone is buying.

You've stumbled into the wrong bar.

Symptom: Your singing voice sounds weak and distorted.
Cause: Beer is too weak, or consumption rate is too low.
Solution: Order several beers at a time.

Symptom: Floor appears blurred.
Cause: Looking through the bottom of an empty bottle.
Solution: Induce someone to buy another round.

Symptom: Your beer seems tasteless and dry.
Cause: Trying to drink from an empty bottle.
Solution: (same as above)

Symptom: The *wall* is suddenly covered with fluorescent lights.
Cause: You have fallen *backwards* onto the floor.
Solution: Ask friends to tie you upright to the bar.

Your mouth contains four or more cigarette butts.

Symptom: Your mouth contains four or more cigarette butts.
Cause: You've fallen *forward* onto the floor.
Solution: (same as above)

Symptom: Floor seems to be in front of you, and moving.
Cause: You're being carried out.
Solution: Ask to be carried into another bar.

Symptom: Room seems unusually dark and quiet.
Cause: Bar has closed for the night.
Solution: Sleep on floor, you will awaken when bar reopens.

Symptom: Your nose and hands hurt, but your mind is clear.
Cause: You've been in a fight.
Solution: Apologize to your friends – in case it was with them.

Beer consumption has exceeded your metabolic limit.

Symptom: Walls, floor, ceiling, and friends' faces revolve.
Cause: Beer consumption has exceeded your metabolic limit.
Solution: Cover your mouth until rotation ceases.

Symptom: You can't remember the words to the songs.
Cause: Your beer and your consumption rate are *just right!*
Solution: Order another round! Play the air guitar!

If a warrior's beer consumption grossly exceeds his metabolic limit, it's not a big problem. The floor is usually a convenient and socially accepted refuge. And, from a safety standpoint, no one has *ever* fallen off the floor. Plus, remember that no one can truthfully claim you're drunk if you can lie on the floor without holding on.

On a safari to Africa, someone forgot the corkscrew and we had to live on nothing but food and water for three weeks.

[William C. Dukenfield (1880-1946), American humorist, who was better known by his stage name, W.C. Fields.]

Murphy's Review of Politically Correct Disease

Politically Correct Disease is a social illness. Those afflicted have degenerated into interpersonal cripples, non-achievers, whiny-babies, and misfits. There is no known cure.

Adults who contract Politically Correct Disease constitute a festering sore on the face of society. These hand-wringing parasites have drifted out of touch with reality. Each day they try to pattern their lives around a warm and fuzzy notion of *fairness* and *inclusion*. These purveyors of psycho-babble build nothing, contribute nothing, risk nothing, gain nothing. To the detriment of society they whine about inequality, discrimination, and a plethora of things that offend them. Although the disease can be frustrating to define, it's easy to identify. Consider the following situation:

These purveyors of psycho-babble build nothing, contribute nothing, risk nothing, gain nothing.

Situation: You are walking down the street with your wife and your children. A deranged man, brandishing a huge knife, begins shouting obscenities and running toward you. You have a Smith & Wesson .44 Magnum pistol (think, *Dirty Harry*) in your hand. You have three seconds before the deranged man – screaming "Kill! Kill! Kill!" – reaches you and your family. **What should you do**?

First, Murphy presents ***The Warrior's Response***:

– The Warrior's Response –

Shoot him. If he's still twitching or moaning after he hits the ground, ***shoot him again!***

Now, the pitiful whiny-baby ***Politically Correct Disease Response***:

– Politically Correct Disease Response –

This isn't fair! There isn't enough information to answer the question. Is the deranged man poor or oppressed? Is he a member of a ***minority group?*** Is he financially or culturally handicapped? Has anyone treated him badly or provoked him? Perhaps someone has discriminated against him and, if so, has he been able to cope with his ***feelings of inadequacy?***

Maybe I could use my gun to knock the knife out of the man's hand. Could I do this without hurting his hand? Perhaps I could run fast enough to get away. If so, maybe the man will feel less anger toward only my wife and my children.

If I don't run, and don't resist, maybe the man will be satisfied with killing only me. But, maybe he doesn't intend to kill anyone. Maybe he just wants to **wound** us. Or, maybe he's having a bad day and wants to **scare** us. Could it be a ***fake knife?***

Maybe my appearance and clothing represent an ethnic insult to the man. Should I try to ***apologize?*** Would he think I'm sincere? If not, and if he does plan to kill me, there's a chance I could grab him and hang on. That way, could my wife and children get away while I'm being stabbed to death?

I could drop my gun, and that might make the man feel less threatened. But, if I fired my gun into the air it might induce him to reconsider his **anti-social conduct**. Or, would that create a greater provocation? And to what extent might it endanger low-flying birds that might be on an endangered species list?

This is absolutely unfair! There are too many unknowns! I need time to think! I need to talk this over with my friends, my spiritual advisor, my attorney, and my psychiatrist. They'll help me find a **non-discriminatory** course of action that (1) won't offend anyone, and (2) will be ***fair*** to all concerned.

As opposed to hapless social cripples afflicted with Politically Correct Disease, all American Warriors are red-blooded patriots. They proudly fly the Flag of the United States of America, and at sporting events they *sing* the National Anthem. Plus, they're easy to identify because they're all gun nuts. In their homes you'll find dozens of back issues of *American Rifleman*, *Shotgun News*, *Gun Digest*, and such.

American Warriors watch wars in the Middle East on T.V., and they always wish they could be there to police-up the empty brass. A warrior's wife and girlfriend both think his aura of Hoppe's No. 9 is his favorite after-shave lotion. And when a warrior passes a magazine rack he always checks-out the cover of *Playboy* – but only after reading *Guns & Ammo* page by page.

A warrior's children are life members of the NRA. Each month a warrior spends more money on ammo than on food. Behind his house you'll find more empty .50 caliber ammo cans than they stock at the National Guard Armory. When it comes to knives, rifles, and machineguns, warriors have enough to equip a couple of platoons, and they have bayonets for rifles they don't own yet.

Warriors know that guns and guts made America free, and they intend to keep it that way. They know so-called gun control efforts aren't about *guns*. They're about *control* by liberal pacifist fools whose vision of a socialist wonderland can't possibly exist anywhere in the real world. Warriors know that you have only the rights you're willing to fight for, and they know:

Warriors know that guns and guts made America free, and they intend to keep it that way.

If you don't know your rights, you don't have any.

Warriors and free men don't ask permission to bear arms.

What part of *"shall not be infringed"* do liberals not understand?

An armed man is a *citizen*. An unarmed man is a *subject*.

Criminals love gun control – it makes their job a lot safer.

Dialing "911" is simply government sponsored dial-a-prayer.

A gun in hand is far better than a policeman on the phone.

Each day 87,976,374 gun owners kill nobody except criminals.

Warriors and free men don't ask permission to bear arms.

Any politician who thinks guns cause crime also thinks (1) matches cause arson, and (2) pencils cause misspelled words.

Guns have only two enemies: (1) rust and (2) stupid politicians.

Know guns, know peace and safety. No guns, no peace and safety.

Unlike irrational defeatocrats with Politically Correct Disease, American Warriors are patriots obsessed with freedom, family values, and the work ethic. Many have a framed photograph of Teddy Roosevelt in their homes and live by his ***Daring Greatly*** philosophy:

– Daring Greatly –

It is not the critic who counts, not the man who points out how the strong man stumbled, or where the doer of deeds could have done them better.

Credit belongs to the ***warrior*** who is actually in the arena, the ***warrior*** whose face is marred by dust and sweat and blood, the ***warrior*** who strives valiantly, the ***warrior*** who errs and comes up short again and again. Credit belongs to the ***warrior*** who knows the great enthusiasms, the great devotions, and spends himself in a worthy cause.

Credit belongs to the ***warrior*** who – at the best – knows in the end the triumphs of high achievement. Credit belongs to the ***warrior*** who – at the worst – if he fails, at least fails while Daring Greatly! His place shall never be with the cold and timid souls who know neither defeat nor victory.

[Murphy has substituted "warrior" in places where Teddy Roosevelt wrote "man"]

An example of the mental degeneration of those with Politically Correct Disease took place on 31 December 2005 in Boston, Massachusetts. On that New Year's Eve some 285 invited guests were packed inside the hotel ballroom for the liberal, politically correct, black-tie gala. About a half hour before midnight an infamous left-wing politician of national note walked to the microphone as the band paused between tunes. The politician apparently was not aware that a freelance contributor to *The New York Times* was also an attendee at the gala – and his pocket-tape-recorder was recording.

The politician could have used the words "Merry Christmas" or "Christmas Holidays," but he didn't. He could have wished the crowd of elitists a "Happy New Year," but he didn't. Instead, he showed America his true colors. Later, when listening to the recording, the freelance contributor could scarcely believe his ears. He had witnessed Politically Correct Disease at its worst, so he typed a ***verbatim transcript*** of the politician's remarks:

– Verbatim Transcript –

[sounds of revelry and laughter slowly subside] . . . Friends! Friends! My wife and I thank you – we all thank you – for joining us here in Boston tonight. Yes, I know – it's snowing down in Washington! *[14 second pause while partygoers applaud]*

Here in the commonwealth of Massachusetts my family and I have been celebrating the ***winter solstice holiday***. We hope you are celebrating too – uuuhhh – accompanied by traditions of your choice. ***No discrimination!*** We don't need ***silly spiritual things*** in our winter solstice celebrations. No! These events always must remain non-discriminatory. *[24 second pause while partygoers applaud]*

In thirty minutes it will be midnight! The start of the generally accepted calendar year two-thousand-and-six. ***Don't discriminate!*** Respect those who may be from other cultures and may recognize some other date as – aaahhh – the start of the new year. *[38 second pause while partygoers shout, cheer, and applaud]*

I'm honored to be with you, regardless of the ***secular beliefs*** you may have – or not have. ***Anything is OK with me!*** I don't discriminate because of your race – uuuhhh – your sexual orientation, your age, your cultural background, or your native language. Enjoy the party!

> *[sounds of wild shouts and cheers as the band begins playing the classic 1964 Phil Ochs liberal anthem, Draft Dodger Rag]*

American Warriors, beware! In addition to the godless hordes promoting political correctness, there are many other enemies of the American way of life and liberty. Long ago you took the Oath of Enlistment for enlisted warriors, or the Commissioning Oath for commissioned warriors. If you're still on active duty you know you are sworn to protect America from "all enemies, foreign and domestic." When you leave active duty you turn in your weapons, fill out some paperwork, and become a military veteran. *Yet, you never rescind your oath*. Warriors, your country still depends on you, your country still needs you. *You are still sworn to defend America* from "all enemies, foreign and domestic."

Warriors, your country still depends on you, your country still needs you.

Domestic enemies like the treasonous blowhard pervert captured on audiotape, above, can be masters of deceit. But, make no mistake, they are enemies of all that American Warriors and their families hold dear. One such person was ***Norman M. Thomas*** (1884-1968). In his day he did his best to subvert this country. He was a dedicated socialist and pacifist, and he was editor of the leftist newspaper, *Nation*. He co-founded the American Civil Liberties Union in 1917. Six times he was the Socialist Party candidate for U.S. President (1928, 1932, 1936, 1940, 1944, and 1948). He explained his agenda for eradicating freedom, liberty, and the American Dream:

Domestic enemies like the treasonous blowhard pervert captured on audiotape, above, can be masters of deceit.

– The Socialist Philosophy –

The American people will never knowingly adopt socialism. But, if we can deceive them with liberalism they will eventually adopt every fragment of the socialist program. America will become a socialist nation – without ever knowing how it happened.

Murphy's Hero: "Old Blood and Guts"

Warriors preparing for battle must be motivated, and nobody inspired and motivated warriors better than "Old Blood and Guts," ***LtGen. George S. Patton Jr., USA (1885-1945)***. He took command of the Third U.S. Army, then staged in England, early in 1944. His soldiers were training for the assault on Fortress Europe during World War II, and he would lead them into combat against the German Army, the potent Wehrmacht.

On 5 June 1944, the day before D-Day, Patton addressed his entire Third Army. Impeccably attired, he strode across the elevated platform to the microphone. His knee-high polished riding boots gleamed in the sunlight. His famed .45 caliber Colt Peacemaker, with ivory grips, was strapped to his waist. Without notes and without a podium, Patton began speaking to his beloved Third Army. There will never, never, never, be another speech like it. Peppered with big doses of eloquent profanity, it is Murphy's selection as the all-time best speech for American Warriors:

There will never, never, never, be another speech like it.

– General Patton's Famous Speech –

Be seated. *[short delay while the soldiers sit down]* Men, this stuff that some sources sling around about America wanting out of this war, not wanting to fight, is a crock of [expletive]. Americans love to fight, traditionally. All real Americans love the sting and clash of battle. . . . You are here because you are real men, and real men like to fight.

> Americans love a winner. Americans will not tolerate a loser. Americans despise cowards. Americans play to win all the time.

. . . Americans love a winner. Americans will not tolerate a loser. Americans despise cowards. Americans play to win all the time. I wouldn't give a hoot-in-Hell for a man who lost and laughed.

Death, in time, comes to all men.

. . . Death must not be feared. Death, in time, comes to all men. Yes, every man is scared in his first battle. If he says he's not, he's a [expletive] liar. Yes, some men are scared, but they fight the same as the brave men, or they get the Hell slammed out of them [while] watching men fight who are just as scared as they are.

The real hero is the man who fights even though he is scared. Some men get over their fright in a minute under fire. For some, it takes an hour. For some it takes days. But a real man will never let his fear of death overpower his honor, his sense of duty to his country, and his manhood.

. . . A man must be alert at all times if he expects to stay alive. If you're not alert some German son-of-a-bitch is going to sneak up on you and beat you to death with a sock full of [expletive]. There are four-hundred neatly marked graves somewhere in Sicily, all because one man went to sleep on the job – but they were German graves, because we caught the [German] bastard asleep before his officers did.

An army is a team. It lives, sleeps, eats, and fights as a team. This individual heroic [talk] is pure horse-[expletive]. The bilious bastards who write that kind of stuff for the *Saturday Evening Post* don't know any more about real fighting under fire than they do about [expletive]. We have the best food, the best equipment,

We don't want any yellow cowards in this army. They should be killed-off like rats!

the best spirit, and the best men in the world. Why, by God, I actually pity those poor sons-of-bitches we're going up against.

We don't want any yellow cowards in this army. They should be killed-off like rats. If not, they will go home after this war and breed more cowards. The brave men will breed more brave men. Kill off the [expletive] cowards and we will have a nation of brave men.

One of the bravest men I ever saw was on top of a telegraph pole in the midst of a furious firefight in Tunisia. I stopped and asked what the Hell he was doing up there at a time like that. He answered, "Fixing the wire, sir." I asked, "Isn't that a little unhealthy right about now?" He answered, "Yes, sir, but the [expletive] wire has to be fixed." Now there was a real man! A real soldier!

Kill off the [expletive] cowards and we will have a nation of brave men.

. . . You should have seen those trucks on that road to Tunisia. Those drivers were magnificent. All day and all night they rolled over those son-of-a-bitching roads, never stopping, never faltering from their course, with shells bursting around them all the time. We got by on good old American guts.

. . . The world is not supposed to know what the Hell happened to me. I'm not supposed to be commanding this army. I'm not even supposed to be here in England. Let the first bastards to find out [that I'm here] be the [expletive] Germans. Some day I want to see them raise up on their [expletive] hind legs and howl, "[Expletive], it's the [expletive] Third Army and that [expletive] son-of-a-bitch Patton."

War is a bloody, killing business. You've got to spill their blood, or they will spill yours. Rip them up the belly. Shoot them in the guts.

We want to get over there [in Germany]. The quicker we clean up that [expletive] mess, the quicker we can take a little jaunt against those [expletive] Japs and clean out their [expletive] nest too, before the [expletive] Marines get all the credit.

Sure, we want to go home. We want this war over with. The quickest way to get it over with is to go get the bastards who started it. The quicker they are whipped, the quicker we can go home. The shortest way home is through Berlin and Tokyo. And when we get to Berlin, I am personally going to shoot that paper-hanging son-of-a-bitch Hitler, just like I'd shoot a snake.

. . . We'll win this war, but we'll win it by fighting and showing the Germans that we've got more guts than they have, or ever will have. We're not just going to shoot the sons-of-bitches, we're going to rip out their living [expletive] guts and use them to grease the treads of our tanks. We're going to murder those lousy Hun [expletive] by the bushel [expletive] basket.

War is a bloody, killing business. You've got to spill their blood, or they will spill yours. So, rip them up the belly. Shoot them in the guts. When shells are hitting all around you and you wipe the dirt off your face, and realize that instead of dirt, it's the blood and guts of what once was your best friend, you'll know what to do. . . . Nobody ever won a war by dying for his country. You win a war by making the ***other*** poor dumb bastard die for ***his*** country.

We are going to twist his balls and kick the living [expletive] out of him all the time.

I don't want to get any messages saying, "I am holding my position." We are not holding a [expletive] thing. Let the Germans do that. We are advancing constantly, and we are not interested in holding onto anything except the enemy's balls. We are going to twist his balls and kick the living [expletive] out of him all the time. Our basic plan of operation is to advance and keep on advancing regardless of whether we have to go over, under, or through the enemy. We will go through him like crap through a goose, like [expletive] through a tin horn.

> . . . There is one great thing you men will be able to say after this war is over and you are home again. You may be thankful that, twenty years from now when you sit by your fireplace with your grandson on your knee – and he asks you what you did in the great World War Two – you won't have to cough, shift him to the other knee, and say, "Well, your granddaddy shoveled [expletive] in Louisiana." No! You can look him straight in the eye and say, "Son, your granddaddy rode with the great Third Army and a son-of-a-[expletive]-bitch named Georgie Patton."

Postscript: Gen. Patton and his Third Army ripped through France and Germany in late 1944 and early 1945, and Patton remained in Europe after Germany surrendered. On 9 December 1945 he was involved in a motor vehicle collision. The accident paralyzed him from the neck down, and he died four days before Christmas. He is buried in the American military cemetery near Hamm, Luxembourg.

The grave of Gen. Patton is marked by a marble cross, center foreground, in an American military cemetery in Luxembourg (photo by the author, Marion Sturkey).

Murphy's Military Superlatives

The ***best*** and the ***worst***. The ***smartest*** and the ***dumbest***. Here are Murphy's Military Superlatives plus crucial Armed Forces information. Wise warriors will learn from these abstracts.

Murphy *chose* these superlatives. He did not create them. Any skeptical reader is encouraged to do his own research to confirm the origin of each superlative Murphy has chosen:

World's Best Literary Tribute to Warriors

William Shakespeare (1564-1616) is often considered the most gifted and imaginative writer the world has ever produced. Murphy selected two lines from Shakespeare's stage-play, *Julius Caesar*, as the ***World's Best Literary Tribute to Warriors***:

Cowards die many times before their deaths;
The valiant never taste of death but once.

= = = = = = = = = = =

Most "Famous Last Words" by a Warrior

LtCol. George A. Custer (1839-1876) commanded the Seventh Cavalry. On 25 June 1876 he found a big Indian village on the Little Big Horn River in Montana territory. Custer dispatched his bugler, John Martin, to ride to Capt. Frederick Benteen and request more ammunition. Then, without waiting, Custer attacked the village.

About 5,200 warriors of the Confederated Sioux Nation were in the village. They chased Custer and his 208 troopers to a nearby ridge and massacred them all. Meanwhile the bugler safely reached Capt. Benteen. He explained Custer's final words to him before he had ridden away, the ***Most "Famous Last Words" by a Warrior***:

Hurry! Hurry! We've caught them napping!

= = = = = = = = = = =

World's Most Effective Draft Notice

The Italian Army began an invasion of Ethiopia, near the horn of Africa, in 1935. The Ethiopian emperor, Haile Selassie (1892-1975), organized a defense of his country. Under his direction the following notice was printed on flyers and posted throughout Ethiopia. Murphy has selected it as the ***World's Most Effective Draft Notice***:

> Everyone will now be mobilized, and all boys old enough to carry a spear will be sent to Addis Ababa. Married men will take their wives to carry food and cook. Those without wives will take any woman without a husband. Women with small babies need not go. The blind, those who can not walk, and those who for any reason can not carry a spear, are exempted. Any [physically fit adult] found at home after the receipt of this order will be hung.

= = = = = = = = = = =

World's Best Philosophy for Warriors

In August 2001 at Fort Worth, Texas, SSgt. Robert Johnson, USA, was speaking informally to a group of high school students. One of the boys asked him to describe a soldier's greatest fear in battle. Johnson's red-blooded reply was right on target. Murphy has selected it as the ***World's Best Philosophy for Warriors***:

> Do not fear the enemy. At the worst, the enemy can only take your life. Instead, a wise warrior fears the so-called news media, for he knows the sniveling media whores may try to steal his honor.

= = = = = = = = = = =

World's Worst Military Prediction

During World War II the U.S. Marines fought and bled their way across the Pacific. Island by island, they rooted-out Japanese defenders. RAdm. Keiji Shibasaki commanded the Japanese garrison on heavily-fortified Tarawa. He knew the Marines were coming, but he also knew the strength of his defenses. He believed his fortress to be impregnable. His written dispatch to Imperial Japanese Army headquarters in Tokyo is Murphy's choice for the ***World's Worst Military Prediction***:

> He believed his fortress to be impregnable.

A million Marines can not take Tarawa in a hundred years.

= = = = = = = = = = =

World's Best Solution for Cowardice

On 5 June 1944, the day before D-Day in World War II, the allies prepared for the invasion of Europe. The flamboyant Gen. George S. Patton Jr., USA, addressed his entire U.S. Third Army. In his polished knee-high riding boots with his .45 caliber Colt Peacemaker strapped to his waist, he strode to the microphone. Without notes and without a podium he spoke for over 25 minutes. His speech included a few choice words about cowardice. These remarks are Murphy's pick for the ***World's Best Solution for Cowardice***:

> We don't want any yellow cowards in this army. They should be killed- off like rats. If not, they will go home after this war and breed more cowards. The brave men will breed more brave men. Kill off the [expletive] cowards and we will have a nation of brave men.

= = = = = = = = = = =

World's Best Military Credo

The U.S. Navy steams Full Speed Ahead, and the U.S. Air Force likes to Aim High. U.S. Army recruiters challenge potential recruits to Be All You Can Be. Famed daredevil French Zouaves chanted their Huzzah, and victorious Japanese warriors shouted Banzai. The international fighter pilot's bar-room

> The volunteer militia had vowed to Remember the Alamo!

cheer invoked gallows humor, *Hurrah for the next man to die!* In Texas the volunteer militia had vowed to *Remember the Alamo!* Worldwide there are many other military credos, but the credo of a U.S. Marine is Murphy's choice for the ***World's Best Military Credo***:

Death Before Dishonor!

= = = = = = = = = = =

Best "Fighting Words" from American Warriors

In the heat of battle or in preparation for battle, "Fighting Words" epitomize the aggressive ethos of American Warriors. They demonstrate the will to win at any cost. From the U.S. Armed Forces, Murphy has picked four statements that exemplify an aggressive spirit and the ***Best "Fighting Words" from American Warriors***:

U.S. Navy: On 5 August 1864 aboard *USS Hartford*, Adm. David G. Farragut, USN, steamed into Mobile Bay. Although the leading ships in his fleet had struck Confederate Navy mines (then called, *torpedoes*), Adm. Farragut shouted:

Damn the torpedoes! Full speed ahead!

U.S. Marine Corps: GySgt. Daniel J. "Dan" Daly, USMC, led U.S. Marines in a bayonet charge against entrenched German Army defenders in Belleau Wood, France, on 6 June 1918. As he ran toward the enemy trenches, Daly shouted to fellow Marines behind him:

Come on, you sons-of-bitches! Do you want to live forever?

U.S. Air Force: In May 1964, Gen. Curtis E. LeMay, USAF, told the Joint Chiefs of Staff how the U.S. Air Force could, and should, deal with the crisis in Indochina:

We'll bomb them back into the Stone Age!

U.S. Army: During the winter "Battle of the Bulge" in World War II, BGen. Anthony McAuliffe, USA, served as acting commander of the surrounded U.S. 101st Airborne Division at Bastogne, Belgium. The Germans dispatched an envoy to

McAuliffe, asking him to surrender. McAuliffe's immortal written reply on 22 December 1944 is quoted below:

To the German Commander:
NUTS!
From the American Commander.

= = = = = = = = = = =

Motivational Shouts for American Warriors

The indigenous American Indians had their *war whoop*. The grey-clad Confederate States Army infantry had its famed *rebel yell*. In combat these shouts were intended to (1) motivate fellow warriors and (2) strike terror within the enemy's ranks.

Today in a barracks environment all gung-ho American Warriors use Motivational Shouts. Warriors use these verbal exclamations to display enthusiasm, agreement, loyalty, and dedication. There are many such shouts, but here is Murphy's pick-of-the-litter for the best ***Motivational Shouts for American Warriors***:

U.S. Army: . . . . Hoo-ah!
U.S. Navy: . . . . Full Speed Ahead!
U.S. Marine Corps: . . Ooo-rah!
U.S. Air Force: . . . . . . Air Power!

= = = = = = = = = = =

Military Academies for American Warriors

The most successful warriors have been properly schooled in military art and military science. The United States maintains three collegiate-level ***Military Academies for American Warriors***:

U.S. Army: . . . . U.S. Military Academy (New York)
U.S. Navy: . . . . U.S. Naval Academy (Maryland)
U.S. Marine Corps: . . U.S. Naval Academy (Maryland)
U.S. Air Force: . . . . . . U.S. Air Force Academy (Colorado)

Combat Philosophy of American Warriors

All warriors share a simple philosophy. They win by *killing the enemy*. Over the years America's warriors have voiced their battle plans in many ways, including the belly-ripping statements Murphy selected as the best ***Combat Philosophy of American Warriors***:

U.S. Army: In England on 5 June 1944, only one day prior to D-Day in World War II, Gen. George S. Patton Jr., USA, addressed the U.S. Third Army. He strode to the microphone and spoke for over 25 minutes. His remarks included:

> Nobody ever won a war by dying for his country. You win a war by making the ***other*** poor dumb bastard die for ***his*** country.

U.S. Navy: Lt. Howell Forgy, USN, was a chaplain aboard the *USS New Orleans* at Pearl Harbor on 7 December 1941. During the Japanese air attack he shouted to the gun crews:

> Praise the Lord, and pass the ammunition!

U.S. Marine Corps: Col. Lewis B. "Chesty" Puller, USMC, commanded the First Marine Regiment at Chosin Reservoir, Korea, in 1950. When told he was cut-off by seven Chinese Army divisions, Puller grinned and told his staff:

> Those poor bastards. They've got us surrounded. Good! Now we can fire in any direction. They won't get away this time!

U.S. Air Force: In 1950, communist hordes attacked American forces in North Korea. MGen. Orvil A. Anderson, USAF, explained how he could prevent Russia from using its nuclear weapons against American troops:

> I can break up Russia's five A-bomb nests in a week. And when I go up to meet Christ, I think I could explain to him that I had saved civilization.

= = = = = = = = = = =

Most Decorated American Warriors

Nitpickers have hit the jackpot. Military *decorations* (for the uninformed, that means *medals*) can not be mathematically quantified. No one can equitably compare a Bronze Star to a Distinguished Flying Cross, or a Purple Heart to the Legion of Merit. Like apples and oranges, they aren't the same thing.

To further muddy the water, military medals are awarded under three circumstances: (1) personal valor in combat, (2) leadership or excellence, and (3) to each warrior in entire military units that have excelled. There's a world of difference between an award for personal valor in combat, and an award to a person whose only claim to fame is that he was *assigned to* a decorated unit.

In making selections, Murphy used simple criteria. The selectee must have received the *majority* of his awards for personal valor in combat (REMFs, administrative pogues, bean counters, chaplain's assistants, and mail clerks are thus excluded). Further, the selectee must have been awarded at least one Purple Heart.

In addition to bullet holes in his body, warriors chosen by Murphy share another trait. Each originally *enlisted* in the Armed Forces. The Navy selectee remained an enlisted warrior throughout his career. The Air Force selectee, who retired as a colonel, began his career as an enlisted Marine. Murphy has selected the following four men as the ***Most Decorated American Warriors***:

U.S. Army: . . . . Maj. Audie L. Murphy (1924-1971)
U.S. Navy: . . . . BMC James E. Williams (1930-1999)
U.S. Marine Corps: . LtGen. Lewis B. "Chesty" Puller (1898-1971)
U.S. Air Force: . . . . . Col. George E. "Bud" Day (1925 ----)

= = = = = = = = = = =

Best Beer-Drinking Song for American Warriors

Wine, wild women, and song are synonymous with the fighting culture of American Warriors. A night of alcoholic revelry the night before battle serves to boost the spirit, promote camaraderie, and ensure restful sleep.

What's the best beer-drinking song for American Warriors? Murphy made a subjective choice. Without question the unanimous choice *should be* the raunchy warrior's rendition of *Let Me Call You Sweetheart*. That's too bad, because only the first line can be printed. The rest of the song is beyond offensive, beyond profane, far beyond obscene. It establishes a brave new frontier for vulgarity, and it's the absolute greatest-ever beer-drinking song! Unfortunately it would burn a hole through any paper it was printed on.

The unanimous choice should be the raunchy warrior's rendition of *Let Me Call You Sweetheart.* That's too bad, because only the first line can be printed. The rest of the song is beyond offensive, beyond profane . . . .

Therefore, Murphy had to compromise. He picked two songs, one for *Infantry Warriors* and one for *Aerial Warriors*. He has listed the primary stanza, followed by the musical bridge (for ignoramuses, *musical bridge* means a refrain with a change of tempo and tune). Here are Murphy's two selections for the Best Beer-Drinking Song for American Warriors. First, the ***Best Beer-Drinking Song for Infantry Warriors***:

– Best Beer-Drinking Song –
for
Infantry Warriors

In peacetime we Regulars are happy,
In peacetime we're willing to serve,
But just when we get a war started,
We'll call out the [expletive] Reserves.

[Musical Bridge]

Call out! Call out!
Call out the [expletive] Reserves, Reserves!
Call out! Call out!
Call out the [expletive] Reserves!

Next, the ***Best Beer-Drinking Song for Aerial Warriors***:

Best Beer-Drinking Song
for
Aerial Warriors

Gory, gory, what a Hell of a way to die!
Gory, gory, what a Hell of a way to die!
Stall! Spin! Crash, burn, and die!
And he'll never fly home again.

[Musical Bridge]

Ten-thousand [expletive] dollars going home to his wife,
Ten-thousand [expletive] dollars in exchange for his life,
More [expletive] money that she's seen in her life,
Just think of all the good [expletive] she can buy!

= = = = = = = = = = =

Birthdays for American Warriors

Murphy has some *revisionist* history to offer. The United States of America did not exist on paper until 4 July 1776. After that, five years of combat were needed to ensure survival of the new nation. But, in 1775 before the United States officially existed, the *Continental Congress* had established the *Continental Army*, the *Continental Navy*, and the *Continental Marines*.

Soon the new U.S. Congress created the U.S. Army, U.S. Navy, and U.S. Marine Corps. Yet, today these three branches of the U.S. Armed Forces consider their official birthday to be the date when their continental forebears were established.

During the early 1900s the Armed Forces had begun utilizing aircraft in combat. In the U.S. Army the aircraft, crews, and support personnel formed the U.S. Army Air Corps (also called *U.S. Army Air Service* and *U.S. Army Air Force*). In 1947 the U.S. Congress finally established an official and independent U.S. Air Force.

For the four primary branches of the U.S. Armed Forces, the official ***Birthdays for American Warriors*** are listed below:

U.S. Army: . . . . 14 June 1775
U.S. Navy: . . . . 13 October 1775
U.S. Marine Corps: . . . 10 November 1775
U.S. Air Force: . . . . 18 September 1947

= = = = = = = = = = =

Best Movies for American Warriors

Another subjective list! Some hand-wringing critics may debate the choices Murphy has made. Many would select *The Green Berets* as the premier U.S. Army film. These advocates could point out that (1) it starred America's foremost screen warrior, John Wayne, that (2) it was the number-one box office draw of its day, and that (3) the patriotic theme song was simultaneously number-one on the musical Hit Parade. Yet, Murphy prefers the *politically impossible* "Old Blood and Guts" and his gung-ho philosophy on offensive warfare. Murphy picks ***Patton***.

Some hand-wringing critics may debate the choices Murphy has made.

Some may question Murphy's selection for the best U.S. Marine Corps film. Fans of John Wayne probably think his immortal *Sands of Iwo Jima* should have gotten the nod. Yet, after watching Jack Webb put his recruits through Marine boot camp, ***The D.I.*** remains Murphy's top choice.

Fans of John Wayne probably think his immortal *Sands of Iwo Jima* should have gotten the nod.

For the best U.S. Navy movie, warrior wannabes probably pick either *Top Gun* or *An Officer and a Gentleman.* Both films are OK if what you want is romantic fantasy. But those looking for a slice of history, butt-kicking naval battles (with lots of original, but colorized, combat footage), and a victorious American underdog will agree with Murphy's selection, ***Midway***.

For the best U.S. Air Force film, some may favor *Twelve O'clock High* and the solemn philosophy of Gen. Frank Savage: "Consider yourselves already dead, once you accept that idea, it won't be so tough." But for a chilling you-are-there ride over Nazi Germany in a shot-to-splinters B-17, fortified with reels of colorized WW II aerial combat film, Murphy picked the best, ***Memphis Belle***. It debuted in 1990 and is the most recent movie on Murphy's list.

In addition to the name of the selected motion picture, Murphy has included (1) the year the film debuted, and (2) the name of the primary starring actor. Here are Murphy's first-place picks for ***Best Movies for American Warriors***:

U.S. Army: .... *Patton*, 1970, George C. Scott
U.S. Navy: .... *Midway*, 1976, Charlton Heston
U.S. Marine Corps: . *The D.I.*, 1957, Jack Webb
U.S. Air Force: ..... *Memphis Belle*, 1990, Matthew Modine

= = = = = = = = = = =

Best Monikers for American Warriors

Murphy knows many nicknames for America's warriors. Navy nicknames, in particular, run the gauntlet from mild and descriptive (such as, *paint-picker*) to crude vulgar appellations that most people never would understand. So, Murphy choose common nicknames that are both (1) descriptive and (2) not overtly obscene. Here are his ***Best Monikers for American Warriors***:

U.S. Army: .... Doggie
U.S. Navy: .... Squid
U.S. Marine Corps: ... Jarhead
U.S. Air Force: .... Zoomie

= = = = = = = = = = =

Military Anthems for American Warriors

The first U.S. Army anthem was *The Caisson Song*. The first U.S. Navy anthem had the same title and tune it retains today, but all the stanzas had different words (about a football game).

Below are today's ***Military Anthems for American Warriors***. For a history of each anthem (and its predecessor, where applicable) plus the words of all stanzas, see the chapter, "Anthems of U.S. Armed Forces and National Anthem":

U.S. Army: .... *The Army Goes Rolling Along*
U.S. Navy: .... *Anchors Aweigh*
U.S. Marine Corps: . *The Marines' Hymn*
U.S. Air Force: ..... *Off We Go into the Wild Blue Yonder*

= = = = = = = = = = =

Official Colors for American Warriors

In the worldwide tradition of military units throughout recorded history, the U.S. Army, U.S. Navy, U.S. Marine Corps, and U.S. Air Force each have two official colors. Here is Murphy's list of these ***Official Colors for American Warriors***:

U.S. Army: .... Black and Gold
U.S. Navy: .... Blue and Gold
U.S. Marine Corps: . Gold and Scarlet
U.S. Air Force: ..... Ultramarine Blue and Golden Yellow

= = = = = = = = = = =

Mottos for American Warriors

A *motto* is a word or phrase often derived from Latin. It's used to describe an intent, motivation, principle, or goal of a group with common interests. Each branch of the U.S. Armed Forces has its own motto, so Murphy lists these four ***Mottos for American Warriors***:

U.S. Army: .... This We'll Defend (the flag)
U.S. Navy: .... Not Self, but Country (unofficial)
U.S. Marine Corps: . Semper Fidelis (meaning, *always faithful*)
U.S. Air Force: ..... Peace Is Our Profession (unofficial)

= = = = = = = = = = =

Mascots for American Warriors

The mule, goat, and falcon are *unofficial* mascots of the U.S. Military Academy, the U.S. Naval Academy, and the U.S. Air Force Academy, respectively. The English Bulldog *official* mascot of the Marine Corps stems from the teufel-hunden (meaning, *devil dogs*) name with which the German Army labeled U.S. Marines during World War I. Shortly thereafter, Marine recruiting posters featured a caricature of a Marine Corps bulldog chasing a terrified German poodle. The ***Mascots for American Warriors*** are listed below:

U.S. Army: . . . . Mule
U.S. Navy: . . . . Goat
U.S. Marine Corps: . . . . . . English Bulldog
U.S. Air Force: . . . . Falcon

= = = = = = = = = = =

Best Invention for American Warriors

Today, American Warriors dine on MREs (Meals, Ready to Eat). But, back during World War II and for years thereafter, warriors got their cuisine in cans and paper packages known as C-Rations. A warrior needed a compact can-opener to open the cans, so in 1942 the Subsistence Research Laboratory developed the famous ***P-38***.

No one is sure how the P-38 got its name.

Only an inch and a half long, the aluminum P-38 is a simple, lightweight, folding-blade, multipurpose tool. It not only opens cans, but it serves as a knife, screwdriver, or whatever. A small hole allows warriors to thread the P-38 onto their dog-tag chain.

The P-38 still thrives today, decades after C-Rations were relegated to the history books.

No one is sure how the P-38 got its name. Some allege that it stemmed from the 38 punctures required to circumnavigate the top of a C-Ration can. Others say the name is based upon a claim that the nifty new tool worked as fast as the then-state-of-the-art Lockheed P-38 fighter plane.

Regardless of the origin of the name, the P-38 still thrives today, decades after C-Rations were relegated to the history books. The fame of the P-38 stems from the unique blend of ingenuity and creativity of warriors who still use it. It's almost impossible to find a crusty old warrior who doesn't still use his trusty P-38, the ***Best Invention for American Warriors***.

= = = = = = = = = = =

Most "Politically Incorrect" Statement for American Warriors

Geoff Metcalf definitely wasn't bashful. He called events as he saw them. On 16 July 2001 he wrote an article entitled, "Run Jane, Run." From that article, Murphy has selected one sentence as the ***Most "Politically Incorrect" Statement for American Warriors***:

> The feminization of the military has been, and remains, a cancer eating away at the warrior spirit and preparedness of the U.S. Military.

= = = = = = = = = = =

Best Epitaph for an American Warrior

In 480 BC the Spartan Warriors who sacrificed all while holding the pass at Thermopylae earned their epitaph. They each remained eternally "obedient to the laws."

> The tombstones of many American Warriors relate the circumstances of their deaths in battle.

Over 2,000 years later in the United States the tombstones of many American Warriors relate the circumstances of their deaths in battle. Some epitaphs are brief. For PFC Thomas Willis, USA, it reads: "Killed In Action on Normandy Beachhead." Other epitaphs offer more detail. The tombstone for LtCol. Clarence K. Hollingsworth, USA, is inscribed: "He died May 16, 1945, in Belgium, of wounds received in action on April 14, 1945, in Germany."

Some epitaphs contain inspirational wording. 1stLt. William M. Rogers, CSA, was killed-in-action on 18 October 1863, and his tombstone inscription includes: "In defence of Southern Rights he laid upon his Country's Altar a life full of the highest promise, in the triumphs of Faith and in hope of a brighter world."

Some epitaphs contain inspirational wording.

Murphy made a subjective choice. He learned that PFC William Cameron, USMC (H Company, 2nd Battalion, First Marines), had been killed-in-action near Lunga Point on Guadalcanal, Solomon Islands, in the South Pacific in 1942. Murphy selected PFC Cameron's tombstone inscription as the ***Best Epitaph for an American Warrior***:

And when he gets to Heaven,
To Saint Peter, he will tell:
"Another Marine reporting, sir;
I've served my time in Hell."

= = = = = = = = = = =

Most Patriotic Statement by an American Warrior

Capt. Jeremiah A. Denton, USN, was an A-6 attack pilot aboard the *USS Independence*. He was shot down over North Vietnam on 18 July 1965 and jailed in the infamous "Hanoi Hilton." After eight years of imprisonment, Denton was the senior officer in the first group of POWs to be released on 13 February 1973. When the big U.S. Air Force C-141 transport rolled to a stop at Clark AFB in the Philippines, Capt. Denton was the first to exit the aircraft. With dozens of TV cameras rolling, he walked to a waiting microphone and spoke three short

He was shot down over North Vietnam on 18 July 1965 and jailed in the infamous "Hanoi Hilton."

sentences. In view of the circumstances, Murphy has selected his remarks as the ***Most Patriotic Statement by an American Warrior***:

> We are honored to have had the opportunity to serve our Country under difficult circumstances. We are profoundly grateful to our Commander-in-Chief and to our Nation for this day. God bless America!

===========

Best Philosophy on Terrorism from an American Warrior

Terrorists crashed four commercial airliners into The Pentagon and the World Trade Center on 11 September 2001, killing themselves plus several thousand Americans and foreign nationals. The next day Maj. Bill F. Weaver, USMC, suggested a proper way to deal with such terrorists. Murphy has selected Weaver's proposal as the ***Best Philosophy on Terrorism from an American Warrior***:

> They should be caught, drawn and quartered, decapitated, and their ugly heads [should be] put on pikes in front of the White House.

===========

American Warriors' Somber Reflections on Combat

Warriors who have tasted combat know that war is not glamor and glory. War, although often necessary in the course of world affairs, is misery and exhaustion, horror and hardship. Warfare is obscene beyond mortal description. Murphy suggests that all boys and warrior wannabes contemplate these ***American Warriors' Somber Reflections on Combat***:

U.S. Army: On 12 August 1880, retired Gen. William T. Sherman, USA, who had led many bloody campaigns during the American Civil War, spoke to a patriotic gathering of military veterans and young men. In part, Sherman told the assembled crowd:

> There is many a boy here today who looks on war as all glory. But, boys, war is Hell.

U.S. Navy: On 3 July 1898 during the Spanish-American War, the victorious American warship *USS Texas* steamed past the burning Spanish warship *Vizcaya* at Santiago, Cuba. The commanding officer of the American ship, Capt. John W. Philip, USN, admonished his cheering crew:

> Don't cheer, men. The poor devils are dying.

U.S. Marine Corps: PFC Ira A. Hayes, USMC, was among six men who raised the American flag atop Mount Suribachi, Iwo Jima, on 23 February 1945. He survived the bloodbath. Two months later a non-combatant told Hayes that, for his role in raising the flag, he was a hero. Hayes replied:

> How can I feel like a hero? I hit the beach with two-hundred-and-fifty buddies, and only twenty-seven of us walked off alive.

U.S. Air Force: During the 1960s, Ray Stubbe was a U.S. Navy chaplain who served at surrounded and besieged Khe Sanh, South Vietnam. In his scholarly 1995 tome, *The Final Formation*, he explained that an American plane had been shot down and had crashed on the embattled base. The pilot, who was still alive, was trapped in the mangled and burning wreckage. Stubbe quoted a witness to his death:

> He was burning to death in the plane and couldn't get out. He was [screaming for] someone to tell his wife that he loved her, and for someone to shoot him.

= = = = = = = = = = =

Murphy's World History
--
(Conservatives vs Liberals)

Today, society in America pits honest hard-working people and patriotic ***conservatives*** against godless hordes of lazy sanctimonious whineybaby ***liberals***. Readers wishing to understand this social conundrum must study *Murphy's History of the World:*

– Murphy's History of the World – (Early Human Society)

Early Human Society: Several million years ago the human race originated in Africa and soon spread into Asia, Europe, and the Americas. Early humans banded together into nomadic tribes of hunter-gatherers. In the summer they camped near streams and learned how to catch and eat fish. In the winter they moved into caves and survived by killing and eating deer and other animals.

Invention of Beer: While the men were sitting around the campfire one night – telling ribald jokes, scratching their genitals, gnawing on deer bones, and breaking wind – they invented beer. These men, our primal ancestors, thrived on the newfound beverage. This evolutionary new development would change human society forever.

Men quickly discovered that grain was a necessary ingredient in beer-brewing.

Beginning of Agriculture: The men quickly discovered that grain was a necessary ingredient in beer-brewing, so they began growing vast quantities of wheat and barley. This worthy endeavor marked the beginning of agriculture.

Evolution of Towns and Villages: Neither cans nor glass bottles had been invented then, so transporting beer from the brewery to the beer-drinkers posed a problem. To solve this dilemma, early humans began living in thatched huts next door to the nearest brewery. This marked the evolution of towns and villages.

> Early humans began living in thatched huts next door to the nearest brewery.

Beginning of Government: When not drinking beer, the stronger men ventured into the forests. They killed wild hogs and returned to their village with pork, which gave them barbecue to eat with their beer. Soon these stronger men began making decisions that affected everyone in the village. This centralization of power in the hands of a few marked the beginning of government.

Invention of the Wheel: As the strong men killed more and more wild hogs, the local hog population began to decline. Soon the men had to hunt farther from their village. They built crude sleds upon which to (1) drag kegs of beer to their hunting areas, and (2) drag dead hogs back to their village. Dragging a sled loaded with beer or hogs was hard work, so the hunters put rolling logs under the sleds. This beneficial innovation marked the invention of the wheel.

> The military dictator would sit in the shade with his buddies, drink beer, eat barbecue, and tell the latest dirty-cave-man jokes.

Evolution of Warriors and Warfare: With wheeled sleds the men could travel great distances with their beer and dead hogs. This caused them to encounter similar strong men from faraway villages. These two groups began fighting each other for hunting rights. This marked the evolution of warriors and warfare.

Military Dictatorships: Victorious warriors became emboldened by their military successes. Therefore, at home in his own village, the strongest warrior became the big boss and made all the decisions for everyone. This marked the beginning of military dictatorships, the world's most efficient type of government.

Life in the Village: When he was not bossing everyone around, the military dictator would sit in the shade with his buddies, drink beer, eat barbeque, and tell the latest dirty-cave-man jokes. Whenever the village meat supply ran low he would (1) gather his warrior buddies together, (2) load the sled with kegs of beer, (3) travel to a good hunting area, (4) kill competing hunters, (5) kill a few hogs, and (6) go home. For the dictator and his warrior buddies, life was good!

Discontent in the Village: Meanwhile, the timid and weak villagers became unhappy. They lacked the courage and strength to fight. They lacked the skills and intelligence to hunt. Day and night they got bossed around by the dictator and warriors. They had to do all the menial cleaning, sewing, mending, cooking, hair-dressing, and such. With no voice in village affairs they subsisted on whatever morsels the warriors gave them. When not doing chores they would huddle together, engaging in group hugs (speaking softly to avoid offending anyone), nibbling on berries and nuts, and whimpering about how *unfair* life had become.

For the dictator and his warrior buddies, life was good!

Conservatives: The dictator and warriors provided the food and protected the entire village. They were realists. They knew fighting and hunting were dangerous, but they were willing to risk life and limb for the greater good. They protected the village wimps and weaklings and provided for their every need. These strong vibrant men believed in conserving their productive way of life, so they called themselves ***conservatives***.

Communists: The weak and timid villagers dreamed of a life without bosses or dictators. They dreamed of a communal neverland, an earthly utopia, a commune. People in a commune would share everything, and life would be fair. Every person in a *commune* would be equal. No person would work harder than anyone else. If they could create such a fantasyland *commune*, life would be wonderful, so they called themselves ***communists***.

The Communist Experiment: The communists hatched a plot. One night when the dictator and warriors were all asleep the communists ran away, abandoning their protectors and providers. Deep in the forest they set up a commune where they planned to live their lives in peace and plenty, sharing everything, engaging in group hugs, explaining their innermost feelings, singing folk songs, building self-esteem, and making sure nobody was treated unfairly. But none of them had ever learned how to hunt, and they were too timid to try, *so they all starved to death within three weeks*.

One night when the dictator and warriors were all asleep the communists ran away.

Murphy's Analysis: Thus ended the world's first experiment with communism. Yet, thousands of years later this ridiculous exercise in stupidity would be repeated during the last two centuries of the Second Millennium.

– Communists, Socialists, and Liberals –

Communists in the United States: Until around 1950 in the United States, communists kept a low profile. Working people and responsible Americans despised them. Government had only a few freebie programs for deadbeats in those days, so communists had a hard time feeding at the public trough. Plus, in that era the term *communist* had been linked to bad guys in the Soviet Union, China, North Korea, and Cuba. Deranged communist wusses in America realized they needed a new image, but what should it be?

Socialists in the United States: In Cleveland, Ohio, two communists were waiting in line for their welfare checks when they experienced an epiphany. They believed in *social* fairness, *social* equality, *social* justice, and other such *social* claptrap. So, they decided they should call themselves ***socialists***.

Almost overnight the new moniker caught on, and it sent shockwaves throughout America. Government cranked up a host of giveaway programs for socialist panhandlers: food stamps, heating subsidies, free housing, free health care, free day care, contract set-asides, employment quotas, affirmative action mandates, more medicaid and medicare programs, etc. For the new socialists, life was good! However, by the 1980s the blush on the rose of socialism had begun to fade. Once again a new image was needed.

They believed in *social* fairness, *social* equality, *social* justice, and other such *social* claptrap.

Liberals in the United States: While filling out their home heating subsidy applications in Newark, New Jersey, three socialists whined to each other about the unfairness of life. Once again, an epiphany! They relied upon government to *liberate* them from poverty, to *liberate* them from unfairness, to *liberate* them from the need to work. Therefore, they decided to call themselves ***liberals***.

The new name swept across America like a prairie fire, and government poured more and more money into efforts to pacify the liberal whiners, complainers, and crybabies. The more freebies the government offered, the more freebies liberals demanded. Liberals realized they were too stupid to care for themselves, so they demanded an omnipresent big-brother government that would make all decisions for them and provide for their every need.

The new name swept across America like a prairie fire.

– Core Beliefs –

Conservatives: Conservatives are thinkers and doers. They work for a living and produce a product or service of tangible use to society.

They view life as a challenge and opportunity. Their concerns are *goals* and *achievements*. They believe in personal responsibility. They think each man should enjoy the fruit of his successes and suffer the consequences of his failures. They believe in hard work, personal initiative, low taxes, small government, and traditional American values. Conservatives are honored to help the sick and the infirm, but they expect all others to help themselves. When their country is threatened their response is simple: ***kill all the bad guys***. Their heroes are people like Ronald Reagan and Teddy Roosevelt.

Their heroes are people like Ronald Reagan and Teddy Roosevelt.

Liberals: Liberals are drones and parasites. They believe in hard work – as long as someone else does it. They produce nothing of use or value to society. They consider life to be unfair. Their concerns are *inclusion* and *equality* and *rights* and *entitlements* and other such foolishness. Unable to care for themselves, they want government to cater to their needs and wishes. Liberals want high taxes (on the working people, but not on them), a benevolent government, a socialized lifestyle, and personal unaccountability. When their country is threatened their response is simple: ***peace-at-any-price***. Their heroes are people like Pee Wee Herman and Milli Vanilli. Several out-of-touch idiotic liberal beliefs are listed below:

Their heroes are people like Pee Wee Herman and Milli Vanilli.

The term "inalienable rights" includes free housing and health care.

Government, not parents, should educate children.

Having self-esteem is crucial, but doing something to earn it is not.

AIDS is spread by a lack of government funding.

Business creates oppression, but government creates prosperity.

All genuine compassion lies in government funding.

Art never existed before government funding to promote it.

AIDS is spread by a lack of government funding.

All genuine compassion lies in government funding.

Career criminals should get counseling, not prison sentences.

Career *criminals* should have the same rights as career *police*.

Government has a moral obligation to confiscate its citizens' guns.

Guns in the hands of law-abiding citizens are more dangerous than nuclear weapons in the hands of terrorists.

Terrorists *may* be bad, but the solution is simple – ignore them.

Killing terrorists is wrong. They're merely freedom-fighters with a different perspective.

Things like (1) Christmas manger scenes and (2) praying to God on private property should be banned, but same-sex marriage is OK.

The word *Christmas* must be replaced with "Winter Holidays" to avoid offending atheists, elitists, immigrants, and terrorists.

At Christmas, (1) manger scenes and (2) praying to God on private property should be banned, but same-sex marriage is OK.

The term "African-American" somehow makes perfect sense.

If racism is the problem, Affirmative Action is the solution.

Standardized tests are racist, but quotas and set-asides are not.

Communism will work in America if liberals simply *keep trying*.

Illegal aliens deserve full medicaid and social security benefits.

Modern-day warriors use "get-some" automatic weapons atop an LAV in their quest to find and kill the bad guys – Murphy suggests that such weapons would be ideal tools for dealing with all whineybaby liberals (photo courtesy of U.S. Marine Corps).

No privately owned business can be allowed to make a profit.

No private citizen can be trusted to handle his own money.

The *communist* and *socialist* and *liberal* agenda to destroy America:
- Encourage ethnic-identity and hyphenated Americanism.
- Let immigrants maintain their former language and lifestyle.
- Work toward a fully multi-cultural and multi-lingual society.
- Promote diversity and discourage unity.
- Provide government funding for the "victimization" industry.

– Typical Occupations –

Conservatives: Conservatives are *happy* people who are willing to work for a living. Many own their own companies and hire other working conservatives, thereby enabling society to function.

Liberals: Liberals are *unhappy* bleeding hearts who never work or create anything of value. Most such losers rely upon government public assistance programs and handouts for their subsistence.

<u>Typical Occupations</u>: Typical conservative occupations and liberal occupations are listed below:

Conservative Occupations	**Liberal Occupations**
Lumberjacks and farmers	Self-esteem group therapists
Military warriors	Social workers of all ilks
Construction workers	Family relations advisors
Entrepreneurs	Marriage counselors
Business CEOs	Psychiatrists
Small business owners	Government bureaucrats
Police officers	Wedding consultants
Firefighters	Hairdressing assistants
Big-game hunters	Interior decorating advisors
Plumbers and electricians	Affirmative Action activists
Commercial fishermen	Psychologists
Professional athletes	Soap-opera reviewers

Liberals are the most good-for-nothing spoiled brats in the world. They feel a great indebtedness toward their fellow man – which they propose to pay with someone else's money. Their so-called economic *reasoning* is that a country can tax itself to prosperity. They call for a redistribution of wealth – from working people to them. Their true passion in life is a crusade to (1) get more and more for less and less and, better yet, (2) get something for nothing. These wusses form a ball and chain around the foot of progress. Most wail and whine for nebulous societal reform, and the symbol of their political affiliation is always a braying jackass.

Liberals are the most good-for-nothing spoiled brats in the world. They feel a great indebtedness toward their fellow man – which they propose to pay with someone else's money.

– The Ant and the Grasshopper – (Conservatives vs Liberals)

Aesop (620-560 BC) was a wise sage who lived in Greece 2,600 years ago and told animal stories to children. These famous stories, known today as ***Aesop's Fables***, hold valuable lessons for mankind. Aesop's most insightful story is "The Ant and the Grasshopper." It can be paraphrased in English as follows:

– The Ant and the Grasshopper – ***(Aesop's Original Version)***

The Ant [***a conservative***] worked hard all summer, building his home and gathering food for the coming winter. Meanwhile, the Grasshopper [***a liberal***] fiddled around in the sunshine all summer long, singing, dancing, loafing, and goofing off.

Winter came, and *the hard-working Ant was warm in his home and had plenty to eat*. The Grasshopper had no home, no shelter from the freezing weather, and no food. He begged the Ant to help him. The Ant refused, because the lazy Grasshopper had been unwilling to help himself. Therefore, *the stupid and lazy liberal Grasshopper froze to death in the snow*.

In the United States the sniveling worthless liberals have become a societal blight. These do-nothings constitute a plague on working people. Murphy knows that if the Ant and Grasshopper saga plays out in America today, the following result will be inevitable:

– The Ant and the Grasshopper – ***(Modern American Version)***

The Ant [a ***conservative***] worked hard all summer, building his home and gathering food for the coming winter. Meanwhile, the Grasshopper [a ***liberal***] fiddled around in the sunshine all summer long, singing, dancing, loafing, goofing off – and wasting his welfare checks on lottery tickets and cheap rot-gut wine.

Winter came, and the Ant was warm in his home and had plenty to eat. The stupid Grasshopper had no home, no shelter from the freezing weather, and no food. He begged the Ant to help him. The Ant refused, because the stupid lazy Grasshopper had been unwilling to help himself. A local T.V. crew filmed the poor helpless Grasshopper kneeling in the snow, shivering, weeping, pleading for government assistance. His welfare check had been stolen, he claimed, and he had nothing to eat and no warm place to lay his head. The cruel Ant had refused to help him, he sobbed, although the Ant was living in a large warm home and had plenty to eat.

A local T.V. crew filmed the poor helpless Grasshopper kneeling in the snow, shivering, weeping, pleading for government assistance.

The next day the Grasshopper's plight made headlines nationwide. Jesse and Al showed up in front of the Ant's home and sang "We Shall Overcome" for the cameramen. Discrimination! Injustice! Shame! Jesse piously proclaimed that such abuse must not be tolerated. The Ant had a warm home and plenty of food. Why had he not shared his wealth with the poor oppressed Grasshopper? Had there ever been a more evil and wicked case of blatant discrimination? Was there no justice in America?

Jesse and Al showed up in front of the Ant's home and sang "We Shall Overcome" for the cameramen.

That night – after coordination with cable news networks – Jesse and Al led a march from City Hall to the Ant's home. The news media filmed the candlelight vigil in the snow, prayers for the poor Grasshopper, curses for the Ant, hundreds of chanting protestors, and statements by attorneys from "Social Justice Now."

Thousands of protestors linked arms, waved signs, and shouted obscenities at the Ant.

The next morning the EEOC filed an emergency petition in District Court. The Judge quickly ruled that (1) the Ant was wealthy, (2) the Grasshopper was indigent, and (3) a "redistribution of wealth" was proper. As thousands of protestors linked arms, waved signs, and shouted obscenities at the Ant, social workers and police evicted him from his home and rushed him to a mental hospital for evaluation. Meanwhile, the Judge granted the Grasshopper temporary "free housing" in the Ant's home.

"Criminal neglect" by the Ant had caused the victimized Grasshopper to suffer "irreversible mental anguish" and "psychic trauma."

A team of court-appointed psychiatrists presented their findings to the Court. They testified that "criminal neglect" by the Ant had caused the victimized Grasshopper to suffer "irreversible mental anguish" and "psychic trauma." A brief filed by "Social Justice Now" attorneys – assisted by the EEOC legal staff – stated that (1) prompt compensation to the Grasshopper, and (2) criminal charges against the Ant, were the proper means to rectify this inexcusable injustice and discrimination.

The Judge ordered all the Ant's assets and property to be transferred to the Grasshopper immediately.

Moved to tears, the Judge ordered that all the Ant's assets and property be transferred to the Grasshopper immediately. The Judge also ordered the "wicked and despicable" Ant to be incarcerated in a super-max federal prison without delay.

Wait, here is "*the rest of the story*."

The stupid Liberal Grasshopper: Based upon the Judge's order, the Grasshopper took the Ant's home, food, and bank accounts. The Grasshopper invited his degenerate liberal friends to a lavish party that lasted a month. They ate all the food. They trashed the house. They spent all the money on liquor, drugs, prostitutes, and hundreds of porno movies. After running through all the money, food, liquor, drugs, and such, the Grasshopper's douche-bag buddies deserted him faster than Ted Kennedy fleeing from a Chappaquiddick car mishap.

The stupid liberal Grasshopper was again destitute. While waiting for his food stamps and welfare check, he tried to sell left-over crack cocaine to make ends meet. Days later his withered bullet-riddled corpse was found face down in the snow. The next day he got his last government freebie, a pauper's burial in a potter's field.

While waiting for his food stamps and welfare check, he tried to sell left-over crack cocaine.

<u>The hard-working Conservative Ant</u>: Meanwhile, the Ant spent his time in the prison library studying computer science. When finally paroled he had no home, no money, and no job. Yet, like all conservatives, the Ant had a strong work ethic. He labored 16 hours per day, worked two jobs at Hardee's and McDonald's, rented a modest apartment, and put food on his table. Within months, using his computer skills, the Ant started working multiple shifts at Radio Shack. In a few more months he became the store manager. Each night in his apartment he toiled tirelessly. He soon developed a revolutionary computer operating system that rivaled Microsoft. The Ant quit his Radio Shack job and worked at home 20 hours each day, refining and marketing his new product. Sales began to soar. He hired other hard-working conservatives to join his new company, help it grow, and share the ever-increasing profits. Within two years the conservative Ant became a multi-millionaire.

Like all conservatives, the Ant had a strong work ethic. He labored 16 hours per day.

The Ant had earned and saved millions of dollars. Yet, he built a modest home and stocked it with cheap, but nutritious, food. He now spends each winter in his warm home and has plenty to eat.

He shares his good fortune with the aged, the ill, the infirm, and those unable to help themselves. However, the Ant tells all *Grasshoppers* and all other deadbeat degenerate whineybaby *liberals* to take (1) a hop, (2) a skip, (3) two giant steps, (4) three running jumps, and (5) kiss his anatomy where the sun never shines!

Murphy's Moral: There is a timeless moral to Aesop's tale of the hard-working conservative Ant and the shiftless good-for-nothing liberal Grasshopper. Hard work and commitment bring success, but laziness and dependence bring privation and misery.

Murphy's modern version of this tale also has a moral, and if Aesop were alive today he would burst with pride. *The moral to Aesop's tale, and the moral to Murphy's version, are identical*.

Long may it wave!

Murphy's History of Stupidity

Governments institutionalized ***stupidity*** in Europe during the Thirteenth Century. When a King was having a bad day he often took out his wrath on his staff, his Royal Court. Consequently, to protect their necks from the executioner's axe the Royal Court guys came up with a self-defense strategy. They began the custom of bringing in some local idiot to make the King laugh and feel better. Over time these local *retards* became known as Court Jesters.

Contrary to present-day belief, Court Jesters were not bright or witty, and they enjoyed no elevated social status. They were chosen for their *sheer stupidity.* Many were so retarded that they lacked the power of speech. The Royal Court delighted in subjecting these poor imbeciles to all manner of indignity and torture, the "fun and games" of the day.

Today's image of a Court Jester usually brings to mind a clown attired in checkered tights and wearing a motley hat adorned with bells. That image is loosely based upon fact. The checkered tights were actually scraps of colored cloth the homeless buffoon had pulled around himself to try to keep warm in wintertime. The bells were attached to his hat by the Royal Court. Tinkling of the bells would further befuddle the retard and add to the King's amusement.

Stupidity is a Capital Offense. The sentence is death, there is no appeal, and the executioner does his job without pity.

Unfortunately each Court Jester led an unhappy and short life. The King usually tired of watching the same idiot after a day or two. Records show that the

King then would decree that the poor man be beheaded to (1) put him out of his misery and (2) rid society of the burden of supporting him. Then the Royal Court would find another retard to become the new Court Jester.

Like death and taxes, stupidity has remained with us over the centuries. Today we know that stupidity isn't a sin. Stupid people didn't choose to be stupid. But stupidity can't be cured with money, or through education, or by legislation. In the game of life, stupidity is a Capital Offense. The sentence is death, there is no appeal, and the executioner does his job without pity.

American Warriors, beware! Stupidity has gained a foothold in the world of military technology. When stupid people input their stupid ideas into a military computer, nothing but their stupid ideas will come out. This fancy new computer-enhanced stupidity, coming from an expensive complex machine, is somehow ennobled. Most of the time nobody has the courage to criticize it. Therefore, American Warriors, be forewarned:

Stupidity has gained a foothold in the world of military technology.

Stupidity is the most powerful force in the universe.

Stupidity knows no boundaries of time and place.

Stupidity is the hobgoblin of ignorant little minds.

Stupidity is the world's most effective armor against reason and logic.

Stupidity cramps any conversation.

Never underestimate the power of human stupidity.

Never underestimate the power of stupid people in large groups.

Stupidity is the hobgoblin of ignorant little minds.

There is nothing more frightening than stupidity-in-action.

A stealthy and lethal Los Angeles-class attack submarine prowls the ocean – Murphy has found that stupid people often wind up aboard such craft after enlisting in the Navy in anticipation of enjoying "yachting" and "water sports" (photo courtesy of U.S. Navy).

Artificial computer intelligence is no match for human stupidity.

One can be sincere and well intentioned, and still be stupid.

For those who are afflicted, stupidity is bliss.

You are only young once, but the afflicted are stupid forever.

Stupidity cramps any conversation.

A little stupidity can go a long way.

Stupidity is no excuse – it's the *real thing.*

Stupidity, plus more stupidity, equals sheer stupidity.

Sheer stupidity is far worse than run-of-the-mill stupidity.

Most of what is erroneously attributed to malice should be attributed to sheer stupidity.

Stupidity is not only more deplorable than we imagine, it's more deplorable than we *can* imagine.

A little knowledge is *dangerous*, but a little stupidity is *worse.*

Stupid people can sometimes spot wrong *answers*, but not wrong *questions*.

For stupid people, mistakes are too much fun to make only once.

Stupid people never make the same mistake twice – they make it three, four, five, or six times.

Stupidity, combined with good intentions, is the most disastrous combination on Earth.

Only a stupid person is always at his best.

For stupid people, mistakes are too much fun to make only once.

Stupidity is the mentality that induces a person to continue throwing tomatoes to a Bengal Tiger, hoping the tiger will become a vegetarian.

Being wise means you *do know* what you don't know, but stupid people *don't know* what they don't know.

A little knowledge is *dangerous*, but a little stupidity is *worse*.

Forgetfulness among liars equates to stupidity – to be successful, a good liar needs a good memory.

If five million people say a stupid thing, it's still a stupid thing.

When "stupidity" is a sufficient explanation, resist the temptation to resort to any other.

Usually, stupid is as stupid does.

Yet, if something appears stupid, but works, it's not stupid.

When stupid people find themselves in a hole, they keep digging.

When a delicate jammed machine needs repair, stupid people never force it. They just use a bigger hammer.

Stupid people drink downstream from the herd.

If five million people say a stupid thing, it's still a stupid thing.

When stupid people find themselves in a hole, they keep digging.

Stupid people will believe anything if someone whispers it.

Stupid people try to train their *dogs* to guard their *hamburgers*.

New theories flourish in direct proportion to their stupidity.

Continuing Education is expensive, but so is *Continuing Stupidity*.

Light travels faster than sound. That explains why stupid people look normal – until they start talking.

From time to time, stupid people try to improve their lot in life by reading. Murphy commends them for their attempt to better their lives, but he knows their efforts never succeed. According to U.S. Library of Congress statistical data, the type of book usually selected by stupid people requires crayons. In a few instances, pencils are needed to connect the dots.

A few stupid people select books which can actually be *read*. Using data from American Literary Research (ALR) files, Murphy identified the most popular such books. He noted a common thread running through each of them. Each is ***extremely short***. Using computerized ALR data files, Murphy has compiled a listing of the 12 most popular ***Short Books for Stupid People***:

> The type of book usually selected by stupid people requires crayons. In a few instances, pencils are needed to connect the dots.

– Murphy's List – of Short Books for Stupid People

1. Fighter Aces of the Iraqi Air Force
2. Morality, Ethics, and the Clinton Presidency
3. Brain Teasers for U.S. Marines
4. Carrier Landings and the B-2 Stealth Bomber
5. Rappelling for Sailors
6. Diplomacy & Courtesy: The Life of Gen. George S. Patton
7. Tasty Gourmet Recipes for MREs
8. Beautiful Golf Courses in Vietnam
9. Under Fire in Bosnia: The Memoirs of Hillary Clinton
10. Victory the McNamara Way
11. Airborne Assaults of the Viet Cong
12. How I Invented the Internet: The Al Gore Story

Murphy's Adages for Fools

Fools are unfortunate people who putter through life with their mouths open and their minds shut. Neither medicine nor therapy can blunt the tragic symptoms, and there is no known cure. The affliction is a life-long curse.

A ***fool*** and a stupid person are not the same thing. Simply stated, a stupid person was intellectually shortchanged at birth. On the other hand, a fool may be smart. His mind and his mouth may function properly – but not at the same time.

> A fool may be smart. His mind and his mouth may function properly – but not at the same time.

Because of the damage they can do, fools have a special day of recognition, ***April Fools' Day***. Until the Sixteenth Century the world celebrated the start of each year on the first day of April. But in 1582, Pope Gregory XIII (1502-1585) scrapped the Julian calendar and introduced his new Gregorian calendar to the Christian world. On the new calendar, each new year began on the first day of January.

The monarchs of Europe soon adopted the Gregorian calendar, and citizens began celebrating each new year when the first day of January arrived. Observers of New Year festivities soon spotted a glitch. A few citizens either (1) "did not get the word" or (2) refused to adapt to change. These slow-learners continued to assume that each new year would begin in April, and they celebrated the first day of that month with revelry and dancing.

> Over 400 years later, April Fools' Day is still alive and well.

Society soon came up with a descriptive moniker for these nonconformists and called them "April fools." Because these fools still celebrated the start of each new year on the first day of April, that day gradually became called the "April fools' day" – ***April Fools' Day***.

Today, over 400 years later, ***April Fools' Day*** is still alive and well. The first day of April is the day to recognize all fools who are mentally out-of-step with society. According to tradition, this is the day to play harmless pranks, the day to trick someone into accepting a silly premise that's obviously false.

Yet, despite the harmless fun on April Fools' Day, the presence of fools in society is no laughing matter. ***Murphy's Analogy about Arguing with Fools*** should be heeded by all professional warriors:

– Murphy's Analogy – about Arguing with Fools

If there are nine fools in the ring, a wise man could jump into their midst and start reciting Shakespeare. Yet, to onlookers, the wise man would appear to be merely the tenth fool.

Warriors must be wary of fools, for they are dangerous. Their minds and their mouths are irrevocably out of sync, and there is no limit to the damage and misery they can cause. Fools can't be cured. They can't be helped. They reject logic. They won't accept suggestions. No one can reason with them. ***Murphy Adages for Fools*** apply to all fools worldwide:

– Murphy's Adages for Fools –

Nothing is foolproof to a talented fool.

To a fool, delusions are functional reality.

Fools often get lost in thought – it's unfamiliar territory.

Fools often get lost in thought – it's unfamiliar territory.

Fools laugh last, but only because they think the slowest.

An ancient P-13 Viper, an antiquated and unreliable fabric-covered flying machine in which romantic fools daydream of engaging in air-to-air mortal combat with modern-day stealthy supersonic fighters (photo courtesy of U.S. Air Force).

The problem with fools and their gene pool – no lifeguard.

Fools rush in where fools usually rush in.

There's no fool like an old fool – you can't beat experience.

Fools consistently *pass up* good opportunities to *shut up*.

Fools rarely resist good opportunities to say nothing.

Only a potential fool will argue with a known fool.

A complete fool causes more damage than a run-of-the-mill fool.

If a run-of-the-mill fool argues with a complete fool, onlookers can't determine who is who.

If a run-of-the-mill fool argues with a complete fool, he must do so on the mental level of the complete fool.

Those tempted to rave at fools run the risk of becoming one.

Only a complete fool will slap a man who's chewing tobacco.

Run-of-the-mill fools kick *old* cow-chips. Complete fools kick *fresh* cow-chips on a hot day.

Only fools wrestle with pigs and argue with idiots.

A hungry fool will eat *several* boxes of prunes – and smile.

Most fools deserve each other.

Those tempted to rave at fools run the risk of becoming one.

Wise men learn from fools, but fools never learn from wise men.

Wise men learn from the mistakes of others. Fools insist on making their own mistakes.

Wise men simplify *complex things*. Fools complicate *simple things*.

Wise men learn from the mistakes of others. Fools insist on making their own mistakes.

Wise men learn by (1) reading, (2) observing, and (3) listening. Fools learn by peeing on the electric fence for themselves.

Smoking a pipe gives a wise man time to think, but it gives a fool merely something to grit his teeth on.

Controversy tends to equalize fools and wise men (unfortunately, this is a fool's one great advantage).

The great irony of life is that fools speak louder than wise men.

The great irony of life is that fools speak louder than wise men.

Most men act like a fool for five minutes each day. Wise men have learned not to exceed the limit.

A fool always wants to be right, for he knows he's not capable of changing his mind.

An M1A1 Abrams Main Battle Tank – which weighs roughly 60 tons and can start moving without warning at any time, day or night – under the front of which several now-deceased persons decided to take a short nap (photo courtesy of U.S. Marine Corps).

Only fools and the dead aren't open to new ideas.

Cynicism is a fool's substitute for intelligence.

Now and then, to err is human. But fools run out of erasers before they run out of pencils.

An argument between fools is an exchange of ignorance.

Now and then, to err is human. But fools run out of erasers before they run out of pencils.

A fool goes through life with his (1) mouth open and (2) mind shut.

Often, it is ethically wrong to allow fools to keep their money.

One way or another, a fool and his money are soon parted.

A fool's version of reason is called prejudice.

Only fools try to stop the march of time.

Fools chatter and chatter when they have nothing to say.

Fools would chatter less if they listened to themselves.

If you don't say a foolish thing, nobody can truthfully quote you, but fools have yet to figure this out.

A talented fool will usually avoid small errors while sweeping on to his grand fallacy.

Fools start out with absolutely nothing, and they still have it.

All kookies aren't found in a jar.

Only fools argue with institutions that buy ink by the barrel.

Fools would chatter less if they listened to themselves.

If there's a wrong thing to say, a fool will inevitably say it.

A drunken fool's words are his sober thoughts.

By chance, even fools are right now and then – but they *think* they're right all the time.

Fools start out with absolutely nothing, and they still have it.

In the battle between fools and the world, the odds favor the world.

A departing fool will fling himself from the room, fling himself upon his horse, and madly ride off in all directions.

Most fools would give their right arm to be ambidextrous.

To a fool, two wrongs are only the beginning.

When there is no solution to a complex problem, a total stranger will step forward and try to take charge – and he will be a fool.

A drunken fool's words are his sober thoughts.

A fool is a person who, just before his untimely demise, shouted to his friends: "Hey, guys, watch this!"

It takes a boy two decades to evolve into a man. But if he's also a fool, it shows in two minutes.

Never approach (1) a horse from the rear, (2) a goat from the front, or (3) a fool from any direction.

It takes a boy two decades to evolve into a man. But if he's also a fool, it shows in two minutes.

However big the fool may be, there is always a bigger fool who admires him.

It's not possible to make any complex system foolproof, because all fools are deceptively ingenious.

So-called "foolproof" systems never take into account the inherent ingenuity of fools.

Ingenious fools can break anything, up to and including anvils.

– Murphy's Warning –

Warriors, beware! There are well-dressed foolish ideas, just as there are well-dressed fools.

Murphy's Guide to Politicians and Government

In the beginning all was well. No politicians! No governments! Anarchy reigned throughout the world, and professional warriors controlled society. Life was good!

Then, sadly, the world began to change. When he wrote *Republic*, Plato (427-347 BC) theorized that the world should be a socialist utopia. He wanted society to be governed by civilians who would be "more qualified than warriors" to handle functions of state. A few decades later in a fit of sheer lunacy, Aristotle (384-322 BC) dreamed up the concept of democracy. He envisioned a world wherein society's rabble, the ignorant masses, would elect a civil ruler.

Warriors of that era should have lopped-off the heads of Plato and Aristotle to squelch the silly democratic and socialist movements.

Warriors of that era should have lopped-off the heads of Plato and Aristotle to squelch the silly democratic and socialist movements in the bud. Tragically they failed to do so, and the world has been on a fast track to disaster ever since.

Today we are plagued with civil control of government. Civil control begets ***elected officials,*** the most spineless, sleazy, amoral life forms on planet Earth. They love to refer to themselves as *public servants*, although they never actually serve the public.

Politics is defined as the process by which these sanctimonious elected officials are chosen. All elected officials have only one goal. They want to get reelected and *remain* elected officials. They are known as ***politicians***, although (1) pervert, (2) parasite, (3) liar, (4) societal whore, and (5) pantywaist would be far more accurate.

Politics is defined as the process by which these sanctimonious elected officials are chosen.

Politicians created the concept of civil government to maintain power and control. As long as they remain elected officials they can continue enriching themselves at the expense of society. Until the world gets so bad that warriors have to rise up and exterminate all politicians, we are stuck with them. To aid professional warriors, Murphy offers these ***Eternal Truths about Politicians***:

No politician can afford to be just a *little bit* crooked.

No politician can afford to be just a *little bit* crooked.

No man's life, liberty, or property will be safe until all politicians are exterminated.

The only difference between a (1) politician and a (2) snail is that a snail leaves his slime behind.

Public Office is the final refuge of the incompetent.

To succeed in politics, one must rise above his principles – if any.

Politicians are *sneaks* in the grass.

For politicians, hypocrisy is the Vaseline of social intercourse.

The only correct way to look at a politician is down.

– How People Degenerate – into Politicians

Those who can – ***do***.
Those who can't do – ***assist***.
Those who can't assist – ***teach***.
Those who can't teach – ***administer***.
Those who can't administer – ***consult***.
Those who can't consult – ***BECOME POLITICIANS***.

A big and lethal M198 155mm howitzer, Murphy's tool of choice for dealing with large groups of parasitic blood-sucking politicians (photo courtesy of U.S. Marine Corps).

The territory in a politician's rhetoric is mined with equivocation.

All politicians have one simple goal in life – getting reelected.

A politician's ethics are as temporary as today's newspaper.

Victory goes to the rich politician who spends enough money to convince the poor that he's on their side.

Whenever a virtuous and worthy cause of the people is entrusted to politicians, it's forever lost.

Government bureaucrats defend the status quo long after the quo has lost its status.

Taxes are contributions which don't benefit the taxed.

When government bureaucrats compromise, the result always will be more expensive than either of the original proposals.

It's impossible to determine if government bureaucrats are (1) sitting on their hands, or (2) covering their butts, or (3) both.

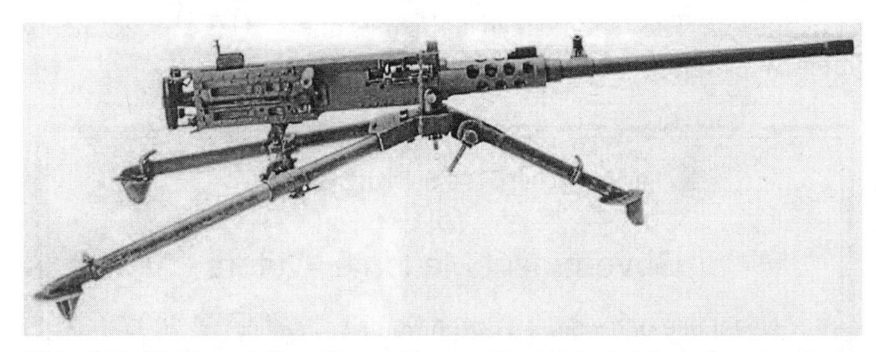

The reliable "Ma Deuce" .50 caliber machinegun, Murphy's tool of choice for exterminating small groups of politicians and parasitic liberals (photo courtesy of U.S. Army).

A government bureaucrat's desk is his castle – and his permanent parking space.

The sole purpose of a government bureaucracy is to perpetuate itself.

The sole purpose of a government bureaucracy is to perpetuate itself.

All actions and decisions within a government bureaucracy are for the purpose of keeping the bureaucracy intact.

If there is a hidden way to delay a crucial decision, an efficient government bureaucrat will find it.

Regardless of need or budget restraints, government will continue to grow at a rate of 14 percent per year.

A taxpayer is someone who never took a Civil Service exam, but still works for the government.

Taxes are contributions which don't benefit the taxed.

Government expands to absorb all tax revenue – and then some.

Regardless of need or budget restraints, government will continue to grow at a rate of 14 percent per year.

Proliferation of new *laws* begets proliferation of new *loopholes*.

Those who want to understand government shouldn't try to read the Constitution. They should read all listings under "U.S. Government" in the Washington phone book.

– Murphy's Rules – for Government Mandate Writers

1. Never use one word if a dozen words will suffice.
2. If the mandate still can be understood, it isn't finished.
3. Mandates must be incomprehensible to the governed.
4. Mandate specifications must be expressed in the least understood terms, such as "furlongs per fortnight."
5. Mandates must protect the mandate writer's job.
6. Mandates must create more mandate-writing government jobs.
7. Mandates must be like an elephant – which is merely a mouse that was built to government specifications.

Congress, when in session, is as dangerous as a drunken homicidal teenage maniac with a chain saw.

If Congress is in session and there's no law now, there soon will be.

The quantity of new legislation is ***directly*** proportional to the degree of mindless media clamor that prompted it.

In any government project, the extent of public benefit is ***inversely*** proportional to the project cost.

In government projects, the degree of accomplishment is ***inversely*** proportional to the amount of paper used.

In government, bad procedures are supplemented, never repealed.

In government, neurosis is communicable.

In government, bad regulation begets worse regulation.

In government, common sense begets nonsense.

In government, growth breeds complexity, which breeds decay.

In government, neurosis is communicable.

No government project will be on schedule or within budget.

The one thing "public servants" never do is serve the public.

– Murphy's Malady –

Whenever government does something that could have been done differently, if always would have been better if it had.

Murphy's Laws of Lust, Sex, and Seduction

Warriors, beware! All the good ones are taken. And if a good one isn't taken, there's always a good reason.

Centuries ago the world of professional warriors remained the exclusive province of men. But in this modern enlightened age there are gung-ho professional women warriors too. Yet, when Murphy speaks of women warriors he means ***real women warriors*** – not the sniveling insecure feminist weaklings and crybabies.

Although ***real women warriors*** are as lethal in combat as their male counterparts, men and women aren't the same species. For the uneducated, Murphy explains the social ethos of the two sexes. First, Murphy presents the ***Male Warrior Social Ethos***:

– Male Warrior Social Ethos –

It's great to be a man! At conception you had a fifty-fifty chance, but you made it. Now the entire world is your private urinal. Your wrinkles give you character. You only have to shave from the neck up. And when you drop by to visit your Male Warrior friends, you bring along a case of beer, not some cutesy little wrapped gift.

Unlike women, you know that two pairs of shoes is more than enough. You can get a three-pack of underwear on-base for less than five bucks, and hair stylists and dry cleaners will never try to rob you blind. If you happen to stumble into the dark and evil snakepit of marriage, at least you get to keep your last name. You're a fighter, a brawler, a killer, a professional population control specialist, a street-legal purveyor of death and destruction, the best America has to offer. You're a gung-ho Male Warrior, and ***men can do anything!***

Next, Murphy presents the ***Woman Warrior Social Ethos***:

– Woman Warrior Social Ethos –

It's great to be a woman! You joined the Armed Forces because you wanted to break things and hurt people. You love the smell of cordite in the morning. Here you enjoy the company of reckless professional warriors with a lethal mindset. You open your own doors. You buy your own beer. You despise the brain-dead *politically correct* mentality of liberal crybabies and bleeding-heart weaklings.

You hold the pathetic hand-wringing feminist whinybabies in contempt. You prefer the heady aura of Hoppe's No. 9 Nitro Powder Solvent over perfume, warpaint over lip gloss, *Guns & Ammo* over the tabloids, combat boots over high heels. Off duty you're a sexual predator without equal, and you're proud of it. You can hold you own and carouse, curse, drink, lie, and fight with the world's best. You're a hard-charging Woman Warrior, and ***women can do anything!***

In love as in war, *nice* warriors finish last. Consequently, when it comes to sex and seduction, lust and love, Murphy has excellent advice. All red-blooded warriors, both men and women, will profit from his time-tested knowledge of erotic arts. In the amorous game of life, he knows the rules. Yet, Murphy cautions the crazed wild-eyed feminist zealots to read the "Warning" on page xiii. His rules are for ***real warriors***, not degenerate liberals. Murphy presents his rules from the ***male point of view***. For women warriors this is no problem. Simply transpose Murphy's gender verbiage and apply his logic. Murphy's eternal truths are divided into six sub-chapters:

1. The Game of Sex and Seduction
2. Warriors, Beware!
3. Living with a Woman
4. Could This Be Love?
5. The Dark Snakepit of Marriage
6. The Woman Warrior's Unique Point of View

– The Game of Sex and Seduction –

In the game of sex and seduction, if you expect to hit the jackpot you've got to drop a few quarters into the slot.

Money can't actually buy love. Nonetheless, it can get you into an excellent bargaining position.

Although money can't actually *buy* love, it can *rent* it.

The best women are free – and worth every penny of it.

Anything worth doing is worth doing for money.

Sex is like air. It's most important when you aren't getting any.

When it comes to the age-old art of seduction, *candy* is dandy, but *liquor* is much quicker.

There are many mechanical devices which increase sexual arousal in women – such as the Mercedes-Benz SL55 AMG convertible.

It can't be premarital sex if you don't intend to get married.

Sex is like air. It's most important when you aren't getting any.

Don't worry, it only seems kinky the first time.

Don't worry. It only seems kinky the first time.

Sex is dirty only if it's done *exactly right*.

No matter how much you get, it's never enough.

Enough is never enough. Take big bites!

No matter how many times you've had it, if it's available, take it, because it won't ever be exactly the same again.

There may be a few – *very few* – things better than sex, but there's nothing exactly like it.

Sow your wild oats, then pray for crop failure.

Never fear! No warrior ever created a baby in *one month* by sleeping with *nine women.*

Frolic only with the best. *Bad girls* make *good company.*

The younger the better. Only wine improves with age.

Sow your wild oats, and then pray for crop failure.

If it's not erotic, it's not interesting, and sex has no calories.

Warriors, it's better to copulate now than never.

Chastity is *not* hereditary, and it can be cured.

Sex *is* hereditary – if your parents never had it, neither will you.

If you never procreate, neither will your children.

Sex appeal is 10 percent what you actually have, and 90 percent what a woman *thinks* you have.

A man should always tell a woman that she's beautiful, especially if she isn't.

Bad girls make *good company.*

The more beautiful a woman is, the easier it is to leave her with no hard feelings.

Nonetheless, all women are beautiful when the lights go out.

If she feels good, do it – you're only as old as the woman you feel.

Sex is one of the nine reasons for reincarnation. The other eight reasons aren't important.

What matters is not the length of the wand, but the magic.

A vasectomy is "never having to say you're sorry."

A dress with a long zipper in the back is the most reliable harbinger of an amorous evening.

Formula for an amorous evening: *scratch her back.*

A woman fond of long zippers will *never* live alone.

Formula for an amorous evening: *scratch her back.*

The main problem with trying to resist temptation is the knowledge that it may never pass your way again.

With a voluptuous and sensuous woman, the only surefire way to get rid of temptation is to yield to it.

Foolish men who abstain from carousing with wild women don't *really* live longer – it just *seems* longer.

Children in the back seat cause few accidents, but accidents in the back seat cause a *lot* of children.

A woman fond of long zippers will *never* live alone.

Three rules: (1) Never eat at a place called Mom's. (2) Never play cards with a man called Doc. (3) Never sleep with a woman crazier than you are.

Whoever coined the term *necking* didn't understand anatomy.

Lust doesn't actually make the world go around, but it makes the ride more exciting.

All women should use what Mother Nature gave them before Father Time takes it away.

The game of seduction is never called off on account of darkness.

Only children, fools, and old reprobates sneer at lust.

– Warriors, Beware! –

When it comes to women, *seeing* can be *deceiving.*

A woman's beauty invites a man's folly.

Love your neighbor – but don't get caught.

Unfortunately, warriors rarely lust after women they can afford.

Love your neighbor – but don't get caught.

When with a woman, a warrior's head is the dupe of his heart.

Romance discriminates against (1) the meek, (2) the timid, and (3) the reticent.

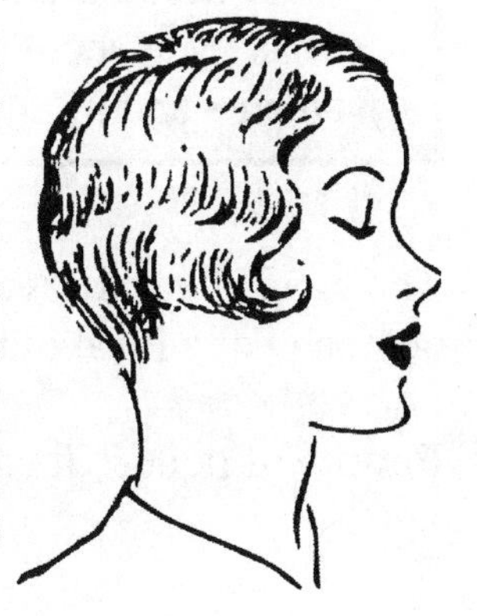

Warriors, your chance of seducing a beautiful woman drops to zero if you're in the company of a better-looking or richer male friend.

The minute a warrior "gets interested" is the same minute a better-looking or richer male warrior will come along.

Avoid frigid women – although they often put on the best show.

A wise warrior never meets a woman, for the first time, in light so dim that he can't read a newspaper.

It's easy to determine when a woman isn't telling a man the truth. He can see her lips moving.

A rabid skunk is better company than a woman with scruples.

If a woman is ever on the verge of saying something smart, she will start by saying, "A man once told me . . . ."

A rabid skunk is better company than a woman with scruples.

Any woman with scruples ought to be shot on sight.

It wasn't the *apple* on the tree, it was the *pair* on the ground, that started all that trouble in The Garden.

The more a warrior learns about women, the more he yearns for the company of a loyal old dog.

All women worry about pregnancy. Unfortunately, they often don't worry at the right time.

The only time a woman ever wishes she was a year older is when she is pregnant.

Beware of liquor. It can make you shoot at women – and miss!

Absence always makes the heart grow fonder – of someone else.

Avoid the fatal "he came, he saw, she conquered" syndrome.

The more a warrior learns about women, the more he yearns for the company of a loyal old dog.

Beware of liquor. It can make you shoot at women – and miss!

Rounds fly downrange as a modern-day warrior demonstrates Murphy's preferred tool and tactic for dealing with women who have scruples (photo courtesy of U.S. Marine Corps).

– Living with a Woman –

Love is blind, but living with a woman is a genuine eye-opener.

Life with a woman is not a journey, but a predicament.

The woman who is *easy to get* is usually *hard to take*.

The woman who is *easy to get* is usually *hard to take*.

The enemy will kill you quickly, but a woman takes her time.

The trouble with women is that they (1) lack the power of conversation, but they (2) retain the power of speech.

Women speak in four tenses: (1) past tense, (2) present tense, (3) future tense, and (4) pretense.

– Murphy's Ten Rules – for Women Who Live with a Man

1. Learn to work the toilet seat. Men need it up, but you need it down. If it's up, put it down. You can do it if you try.
2. If you want something, ask for it. ***Speak English!*** Little hints won't work. Big hints won't work either.
3. Unless your home is on fire, speak only during commercials.
4. If a man asks you what's wrong, and you say "nothing," the conversation is over. ***Stop sniveling*** and learn to live with it.
5. If you have a problem, men offer solutions. If you're looking for sympathy, call your girlfriends.
6. If you think you're fat, maybe you are. ***Never ask!*** Men are smart enough not to answer.
7. You have enough clothes.
8. You have far too many shoes.
9. Try to understand that men only deal with the three basic colors. ***Pumpkin*** is a fruit. It's not a color. ***Peach*** is also a fruit. And whatever ***Mauve*** and ***Beige*** may be, they aren't colors.
10. Pouting, sulking, and crying are blackmail. ***Don't go there!***

All women are "generous to a fault" when it's *their* fault.

When dealing with a woman, a warrior ought to tell her the truth – unless, of course, he is a talented liar.

A woman rarely passes gas in public – she can't keep her mouth shut long enough to build up the required pressure.

Warriors prefer women who wag their tails, not their tongues.

A woman rarely passes gas in public – she can't keep her mouth shut long enough to build up the required pressure.

Women may forgive and forget – but they never *really* forget.

There are two surefire ways to explain complex technological issues to a woman – but neither of them has ever worked.

In the overall scheme of things, sex takes (1) the least amount of time and causes (2) the greatest amount of trouble.

Women who always "sleep like a baby" won't ever have one.

Always be sincere with a woman, even when you have to fake it.

An honorable warrior never hits a woman unless she deserves it.

A man without a woman is like a neck without a pain.

In order of preference, men like (1) dogs, (2) beer, and (3) wide screen TVs. An exceptional woman *may* sneak into the top ten.

Men prefer loyal old dogs because, unlike women, dogs don't whine unless something is *really* wrong.

Men prefer loyal old dogs because, unlike women, dogs don't whine unless something is *really* wrong.

The primary difference between a (1) *whining woman* at the back door and a (2) *whining dog* at the back door is that, if you let them both inside, the dog will quit whining.

– Could This Be Love? –

A warrior can't buy love. Nonetheless, he will pay heavily for it.

Love is the triumph of imagination over intelligence.

Love is the cruel delusion that one woman differs from another.

There is no difference between a (1) wise man and a (2) complete fool if they both fall in love.

Love is an *ideal* thing. Lust is a *real* thing. Confusing the two usually brings disaster.

Love is an *ideal* thing.
Lust is a *real* thing.
Confusing the two
usually brings disaster.

Never confuse love and lust. Love is a matter of *chemistry*. Lust is a matter of *physics*.

Love and lust, although different, both require an accomplice.

A warrior can be happy with any woman if he doesn't love her.

Warriors who *fall* in love usually have to *climb* out.

Love is blind. If Jack's in love, he's a poor judge of Jill's beauty.

When love wears *thin*, faults grow *thick*.

Those who are sensible about love are incapable of it.

Platonic love exists only from the neck up.

If a warrior won't lie to a woman in love, he has little consideration for her feelings.

In many instances, religion has done lust a big favor by proclaiming it to be a sin.

If a warrior feels nauseous and tingly all over, he's either (1) in love, or he (2) has smallpox. Both have dire consequences.

– The Dark Snakepit of Marriage –

Centuries ago a marital union was simple. If a man and a woman consensually lived together, they were "married." The first trouble started in the English Parliament in 1753. Lord Hardwicke's Act required a (1) "banns" or "license" and a (2) ceremony by the Church of England before a man and a woman could live together. No problem, most hot-to-trot couples simply fled to nearby Scotland where only "mutual consent" was required for cohabitation. But, unfortunately, Lord Hardwicke and his goofy mandate had signaled the start of ***government intervention*** in marriage.

In the new United States things started off well. In 1809 the New York legislature was the first to endorse "common law" marriage. Thereafter, mutual consent was sufficient to create a legal marital union. But, tragically, the Twentieth Century brought disaster. Married men and women became entangled in an inextricable web of legal obligations. These entanglements transformed the civil institution of marriage into a ***penal institution***. Those trying to escape faced property battles, custody disputes, the albatross of alimony, and civil sanctions without an end. Therefore, Murphy offers his invaluable advice and unique wisdom to warriors who may be tempted to get married:

These entanglements transformed the civil institution of marriage into a ***penal institution***.

If a man and a woman consensually lived together, they were "married."

A bachelor looks before he leaps – and then doesn't leap.

The only happy people are (1) married women and (2) bachelors.

Bachelors know more about women than married men. Otherwise, bachelors would be married men too.

It's been said that a warrior is *incomplete* until he gets married. After that, of course, he's *finished.*

A warrior will never experience true joy and happiness until he gets married – but then it's too late.

A groom is a man with his last chance for happiness behind him.

Love is "never having to say you're sorry." Marriage is never having a chance to say anything.

After a warrior marries *Miss Right* he finds out that her first name is *Always.*

A married man's home is his hassle.

A warrior may be a fool and not know it – unless he's married.

A husband is all that's left of a married man – after his spirit dies.

Unfortunately, no marriage certificate has an expiration date.

Sooner or later every married man will believe in Hell – on Earth.

All marriages are bliss for about a month. It's the living together afterward that's Hell – on Earth.

Marriage is a lottery, but when you lose, you must keep the ticket.

Successful marriage, if there were such a thing, would require (1) a blind wife and (2) a deaf husband.

A groom is a man with his last chance for happiness behind him.

The only happy people are (1) married women and (3) bachelors.

With a woman, marriage is only a brief suspension of hostilities.

For warriors, the true definition of marriage is *slavery.*

For a warrior, a golden ring on the third finger is akin to an iron ring through the nose.

For a warrior, marriage is like taking a bath. It usually isn't so hot once you get accustomed to it.

Marriage is the process by which the grocer gets the florist's account.

Warriors marry the woman who is *most available* when they are *most vulnerable*.

Warriors marry the woman who is *most available* when they are *most vulnerable*.

Before his liver fails, a warrior can never drink enough of his mother-in-law's booze to get even.

– Murphy's Marital Factoid –

The now-famous U.S. Government "Health and Welfare" study in 1998 revealed, in part, that 94 percent of all married men, (1) age 35 and over, who had been (2) married for at least five years, did not want their mother-in-law to live within a day's driving distance.

Scientists have discovered a food that decreases a woman's sex drive by 93 percent. It's called "Wedding Cake."

An ex-girlfriend is a *memory*, but an ex-wife is *forever*.

A wife never forgets where she went on her honeymoon. A man may forget *where*, but never *why*.

An ex-girlfriend is a *memory*, but an ex-wife is *forever*.

Lust is a *quest*, seduction is a *conquest*, divorce is an *inquest*.

Love is grand. Divorce is a hundred grand.

In marriage, a *horrible ending* is always much better than *horrors without end.*

– The Woman Warrior's – Unique Point of View

Smart women warriors know that men – like coffee and chocolate – are much better if they're rich.

Women warriors prefer their men to have something *tender* about them – as long as it's *legal tender*.

"All work and no play" makes Jack a dull boy, but it makes Jill an exceptionally rich widow.

A man should never try to out-stubborn a woman.

A woman warrior's greatest labor saving device is a rich man.

Behind every rich husband is a woman – with nothing to wear.

A woman warrior knows her man should never argue with her when she's tired – or rested.

There are always two sides to every argument – unless one of the parties to the argument is a woman.

If a man argues with a woman, and it turns out that he's right, he should apologize immediately.

If a man argues with a woman, and it turns out that he's right, he should apologize immediately.

A man should never try to out-stubborn a woman.

Women pick their men – *to pieces*.

If a woman ever wants her man's opinion, she gives it to him.

Today, the things a woman *adores* about her new man will be the same things that she'll *despise* within two weeks.

A smart woman never appeals to her man's *better nature*. She knows most men don't have one.

Love is a matter of chemistry. That explains why smart women treat their men as toxic waste.

For a woman on the prowl, sex is like the falling snow.

(Ladies, if you don't know the pithy "falling snow" analogy that proves this axiom, too bad, because Murphy isn't telling!)

By the time a woman *finally* understands her man, she's grown tired of listening to him.

– Murphy's Advice –

Women and their cats will do as they please. Men and their dogs may as well relax and get accustomed to it.

Murphy's Rules for Dating a Warrior's Daughter

Teenage boys, listen-up! When you see the father of the girl you wish to date, do your homework. Is he a warrior? Is he a military veteran? If so, Murphy has the rules for your survival.

If you expect to ***stay alive*** while sniffing around at a warrior's home, memorize and obey Murphy's Rules.

You may think the father is merely a balding middle-aged reprobate, but do not be deceived. Yesterday's warriors and military veterans have *been there* and *done that*. When it comes to sex and seduction, love and lust, they know every sneaky little devious trick in your book. If you expect to ***stay alive*** while sniffing around at a warrior's home, memorize and obey Murphy's Rules.

To make it easy for teenage male perverts and degenerates to comply, Murphy wrote his 17 rules from the ***warrior's point of view***. Ignore these rules at your own peril:

– Murphy's Rules –
for
Dating a Warrior's Daughter

Rule 1: If you stop in my driveway and honk your horn, I'll assume you wish to deliver a gift. You won't be picking anything up.

Rule 2: After I've searched you and checked at least two forms of photo-identification, you may stand-at-ease by my front porch.

Rule 3: Don't start jabbering about sports, politics, cars, and such. Unless I tell you to talk, ***keep your mouth shut***.

Don't start jabbering about sports, politics, cars, and such. Unless I tell you to talk, keep your mouth shut.

Rule 4: If more than an hour passes, just stand there. My daughter usually takes several hours to put on her makeup.

Rule 5: Keep your hands out of your pockets. If you jingle your car keys, I may think bad guys are in-the-wire if my PTSD has flared up.

Rule 6: While you wait you may wish to do something useful. If so, raise your hand, speak, and request permission to wash my truck.

Rule 7: Don't wander around. My Pit Bulls are always on the lookout for fresh new meat.

Never lie to me! When it comes to my little girl, you have ***one chance*** to tell me the truth, the whole truth, and nothing but the truth.

Rule 8: ***Never lie to me!*** When it comes to my little girl, you have ***one chance*** to tell me the truth, the whole truth, and nothing else but the truth. I have (1) no scruples, (2) a house full of automatic weapons, (3) enough ammo for a Marine rifle company, (4) a back-hoe, and (5) forty-six wooded acres behind the house. ***Don't tempt me!***

Rule 9: When my daughter is ready, you're free to go. As you leave I'll ask when you plan to bring my daughter back to my home. Your reply should be: "***Sir, whenever you say***."

Rule 10: You must ***never*** take my daughter to:
- Places where there are no nuns or police officers within sight.
- Places where there are beds, sofas, recliners, and such.
- Places where there is:
  (1) Music or singing.
  (2) Dancing or hand-holding.
  (3) Happiness or frolicking.

Rule 11: You ***may*** take my daughter to:
- Nursing homes, churches, government offices, and hospitals.
- Military parades and patriotic functions.
- Auto racing events and gun shows.
- Movies, but only ***if*** two conditions are met:
  (1) The movie must have no romantic themes.
  (2) You and my daughter must not sit in adjacent seats.

Rule 12: Never, never touch my daughter. If I see your hands on my daughter, I will remove them. It will be painful.

Rule 13: If you ever make my daughter cry, I'll make you cry a lot more before I kill you.

You probably know that sexual contact without using "protection" can kill you. If you even ***daydream*** about sexual contact with my daughter, I ***am*** the protection, and I ***will*** kill you.

If you ever make my daughter cry, I'll make you cry a lot more before I kill you.

Rule 14: You likely know that sexual contact without using "protection" can kill you. If you even ***daydream*** about sexual contact with my daughter, I ***am*** the protection, and I ***will*** kill you.

Rule 15: While you're away with my little girl, the voices in my head tell me to lock-and-load. I do what the voices tell me to do.

<u>Rule 16</u>: When you return to my driveway, keep both hands on the wheel. Keep your filthy eyes off my daughter. Don't forget about *<u>Rule 12</u>*, and don't even think about exiting your vehicle.

<u>Rule 17</u>: After my daughter has safely entered my home, ***<u>drive away immediately</u>***. The camouflaged old warrior lurking behind the azalea bushes is me.

> Keep your filthy eyes off my daughter.

Murphy's Guide to the Chinese Zodiac for Warriors

Warriors, are you frustrated by life? Unhappy with your moon sign (Aquarius, Pisces, Scorpio, or whatever)? Unlucky in love? Are you burdened by financial problems? Under a Voodoo Hex? Has some wicked witch cast an evil spell on you? Does there seem to be no way out? If so, don't despair. Murphy suggests that you turn to the Chinese Zodiac.

Science has proven that Astrology is bunk. Conversely, Chinese Zodiac "signs" are based upon the true *cyclic* nature of time, as opposed to the erroneous *linear* concept that is popular in the Western World. Scientists and astronomers know that planets orbit the sun in a repetitive and predictable pattern. The Chinese calendar incorporates this 12 year cosmic cycle. Murphy knows that a warrior's knowledge of this planetary pattern is crucial to happiness and success.

Has some wicked witch cast an evil spell on you?

Chu Yuan-Ghang, the legendary Chinese soothsayer, proved that the *day* of a warrior's birth does not dictate his destiny. What matters is the *year* in which the warrior was born. As scientific studies in Munich, Germany, have recently confirmed, the character and personality traits with which a person is born will be duplicated in people born 12 years thereafter.

Chu Yuan-Ghang, the legendary Chinese soothsayer, proved that the *day* of a warrior's birth does not dictate his destiny.

Twenty-six centuries ago Buddha (563-483 BC) bestowed an ***animal sign*** upon each year in the cycle. Each sign represents one year in the Chinese Zodiac. Murphy has consolidated these gems of wisdom for the benefit of all mankind. Look on the following pages and find the year of your birth. Then, simply read your *sign* to unlock the traits which dictate your future and fate:

Year of the Rat: (for people born during the years 1924, 1936, 1948, 1960, 1972, 1984, 1996, 2008, 2020, 2032, etc.)

People born during a year of the Rat are noted for personal charm, generosity, imagination, and an inclination toward opportunism. They strive hard to achieve their goals. Their ambitions are boundless, and they usually accumulate wealth and many material possessions.

On the other hand, Rat people are perfectionists. When things do not go their way they often fly into quick-tempered fits of rage. Also, to their detriment, they are prone to vicious gossip. This inevitably causes their interpersonal relationships to suffer greatly.

Soulmates: Rat people are most compatible with members of the opposite sex who were born during a year of the Ox or Monkey.

Disaster: To avoid misery, a Rat must avoid all Horse people.

Year of the Ox: (for people born during the years 1925, 1937, 1949, 1961, 1973, 1985, 1997, 2009, 2021, 2033, etc.)

People born during a year of the Ox are talented, smart, articulate, and virtuous. They speak softly, yet eloquently, and inspire trust and confidence in others. Ox people are dependable deep-thinking philosophers who relish the company of like-minded friends.

Unfortunately, Ox people are inclined toward a host of unreasonable fears. They are rarely able to conceal their greatest weakness, a fierce temper and a never-back-down stubborn streak. Their inner battle against anger and conflict saps much of their strength.

Soulmates: Ox people are most compatible with members of the opposite sex who were born during a year of the Rat or Rooster.

Disaster: To skirt disaster, an Ox must never consort with Goats.

Year of the Tiger: (for people born during the years 1926, 1938, 1950, 1962, 1974, 1986, 1998, 2010, 2022, 2034, etc.)

Tiger people are deep thinkers, yet courageous and powerful. They fear neither authority nor foes, and they go boldly into the breach when lesser persons quake in fear and inactivity. Others have great respect for Tigers because of their obvious audacity.

Conversely, Tiger strength is a double-edged sword. Tiger people are suspicious and sometimes paranoid. This causes conflict with those in authority. Also, Tiger people are prone to make snap decisions without benefit of necessary facts, often resulting in poor choices.

Soulmates: Tiger people are most compatible with members of the opposite sex who were born during a year of the Dragon or Dog.

Disaster: Tigers must steer clear of Monkeys at all costs.

Year of the Rabbit: (for people born during the years 1927, 1939, 1951, 1963, 1975, 1987, 1999, 2011, 2023, 2035, etc.)

Rabbit people are intuitive, talented, ambitious, and endowed with refined and exquisite tastes. Although fond of gossip, they manage to remain tactful and kind. Rabbit people seldom lose their temper, and they are always surrounded by a bevy of close friends.

Yet, Rabbits are obsessed with gambling, and they seem to be cursed with perpetual bad luck. They have a knack for choosing the wrong thing at the wrong time. This usually leads to financial disaster, which is why Rabbits often must rely on the generosity of friends.

Soulmates: Rabbit people are most compatible with members of the opposite sex who were born during a year of the Pig or Goat.

Disaster: To avoid certain doom, Rabbits must shun all Roosters.

Year of the Dragon: (for people born during the years 1928, 1940, 1952, 1964, 1976, 1988, 2000, 2012, 2024, 2036, etc.)

Dragons are energetic, excitable, short-tempered, and stubborn. Yet, they are brave and consequently inspire confidence and trust. They always seem to be full of energy and vitality. Dragons are wildly popular with others because of their boundless enthusiasm.

Nonetheless, Dragon people usually turn out to be too softhearted – a trait that gives unscrupulous persons the opportunity to take advantage of them. Dragons desperately try to conceal this inherent weakness, but usually their attempts fail, and they remain vulnerable.

Soulmates: Dragon people are most compatible with members of the opposite sex who were born during a year of the Snake or Tiger.

Disaster: Dragons must never keep company with Dog people.

Modern warriors fight not only on the land, but from the air and sea. This big Arleigh Burke class guided missile destroyer, the USS Momsen from the U.S. Seventh Fleet, prowls the waters of the Indian Ocean in September 2008 (photo courtesy of U.S. Navy).

Year of the Snake: (for people born during the years 1929, 1941, 1953, 1965, 1977, 1989, 2001, 2013, 2025, 2037, etc.)

People born during a year of the Snake are stingy and rarely have to worry about money. They are creative, analytical, goal oriented, and obsessive. They reject the status quo and favor new concepts, theories, and ideas. Although Snakes cultivate many like-minded friends, they rely upon themselves and reject advice from others.

Unfortunately these traits tend to transform a Snake into a technical and social recluse. Snakes are perfectionists, and they become frustrated with those who do not share their values and beliefs.

Soulmates: Snakes are most compatible with members of the opposite sex who were born in a year of the Dragon or Rooster.

Disaster: To circumvent calamity, Snakes must avoid all Pigs.

Year of the Horse: (for people born during the years 1930, 1942, 1954, 1966, 1978, 1990, 2002, 2014, 2026, 2038, etc.)

People born during a year of the Horse are cheerful, perceptive, and skillful with money. They are tenacious about everything in life, especially their daily work. Gregarious and outgoing, they love entertainment and seek pleasure with crowds and associates.

On the other hand, Horses inevitably talk too much, and this has an adverse effect on potential friendships. Worse yet, they have a pronounced weakness for insecure overtures from members of the opposite sex, and often this leads to a lifetime of misery.

Soulmates: Horse people are most compatible with members of the opposite sex who were born during a year of the Tiger, Goat, or Dog.

Disaster: To thwart failure, Horses must shy away from all Rats.

Year of the Goat: (for people born during the years 1931, 1943, 1955, 1967, 1979, 1991, 2003, 2015, 2027, 2039, etc.)

People born during a year of the Goat are elegant and accomplished in the arts. They are passionate about what they do and ideas in which they believe. They enjoy achievements, the best things in life, and the things that money can bring. Consequently, they spend lavishly.

Although Goat people spend *lavishly*, they also spend *foolishly*. This results in extreme financial problems that will plague them forever. Although highly creative, they are usually pessimistic about their lot in life, for their dreams always exceed their financial means.

Soulmates: Goat people are most compatible with members of the opposite sex who were born in a year of the Rabbit, Pig, or Horse.

Disaster: To stave off doom, a Goat must never befriend an Ox.

Year of the Monkey: (for people born during the years 1932, 1944, 1956, 1968, 1980, 1992, 2004, 2016, 2028, 2040, etc.)

Monkeys are the erratic people of the planetary zodiac cycle. They are clever, skillful, flexible, and remarkably inventive. They solve the most complex problems with ease and are blessed with a strong desire to succeed. Their cheerful personality is an invaluable asset.

Nonetheless, Monkey people always want to do everything *right now*. When they can not start immediately they are prone to become discouraged, and they often abandon worthy projects in anger. Their frustration and vacillating nature thwart many of their goals in life.

Soulmates: Monkey people are most compatible with members of the opposite sex who were born during a year of the Dragon or Rat.

Disaster: A wise Monkey person will never dally with a Tiger.

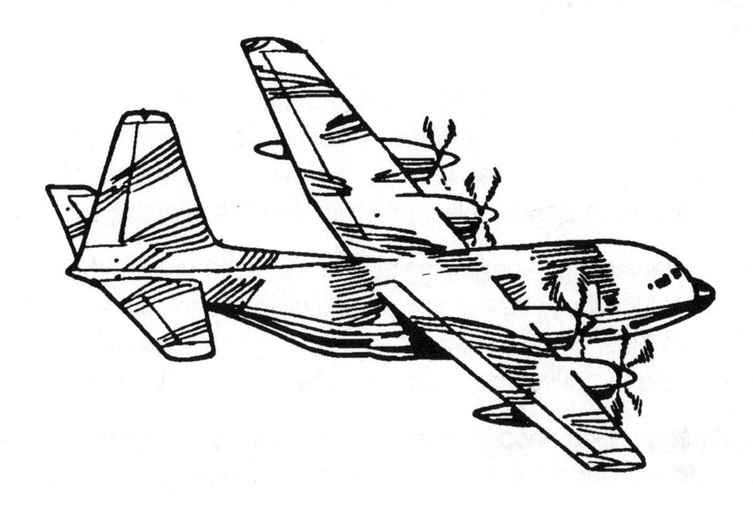

Year of the Rooster: (for people born during the years 1933, 1945, 1957, 1969, 1981, 1993, 2005, 2017, 2029, 2041, etc.)

People born during a year of the Rooster are deep thinkers, capable, and talented. They dare to dream beyond the scope of status quo mentality. True, they often build "castles in the air." Nonetheless, their visions for the future bear fortuitous fruit for all mankind.

Yet, Roosters are eccentric. Their interpersonal relationships suffer because of their irritating habit of insisting on "getting their way." They live most of their lives trying to prove they are superior to others, and lifetime happiness, consequently, almost inevitably eludes them.

Soulmates: Rooster people are most compatible with members of the opposite sex born during a year of the Snake or Ox.

Disaster: For a chance at happiness, Roosters must avoid Rabbits.

Year of the Dog: (for people born during the years 1934, 1946, 1958, 1970, 1982, 1994, 2006, 2018, 2030, 2042, etc.)

People born during a year of the Dog have a refined sense of loyalty and honesty. They never betray a friend. Dog people also inspire confidence in others because they know how to keep secrets. Most of all, friends value Dog people for their lifelong faithfulness.

Notwithstanding the above, Dog people are prone to be stubborn. In social settings they often seem cold and distant, and sometimes they lapse into bitter sarcasm and cynicism. These traits make it hard for Dog people to cultivate new social and business relationships.

Soulmates: Dog people are most compatible with members of the opposite sex who were born during a year of the Horse or Rabbit.

Disaster: To succeed in life, a Dog must keep all Dragons at bay.

Year of the Pig: (for people born during the years 1935, 1947, 1959, 1971, 1983, 1995, 2007, 2019, 2031, 2043, etc.)

People born during a year of the Pig are chivalrous and gallant. Whatever they do, they do with all their strength. Pigs charge straight ahead and attack all challenges and obstacles without delay. They prize loyalty above all else, and their friendships usually last a lifetime.

However, all Pigs are quick-tempered and are inevitably drawn into conflicts and quarrels. No matter how minor the problem, they find a way to argue and make it worse. Their impulsive argumentative nature frequently alienates those engaged in joint endeavors with them.

Soulmates: Pig people are most compatible with members of the opposite sex who were born in a year of the Rabbit or Goat.

Disaster: To avoid peril, a wise Pig never consorts with a Snake.

Murphy's Introduction to Redneck Warriors

The cream of American youth will grow up to become warriors. Most of these professional assassins will be ***Regular Warriors***, but 20 percent will be ***Redneck Warriors***. Neither variant is necessarily good or bad, they're merely *different*. One variant speaks Regular English, and the other speaks Redneck English. Plus, there are many cultural differences between the two species.

It is imperative that each warrior be able to categorize each of his brothers-in-arms as a (1) Regular Warrior or a (2) Redneck Warrior. In combat both variants are equally lethal. However, in a social environment they exhibit a host of differences. If these differences are not thoroughly understood, communication and unit cohesion will suffer. Regular Warriors must learn to appreciate the culture and values of their Redneck Warrior compatriots. Only then can they bond together into a fearsome fighting team.

> In social environments they exhibit a host of differences.

Murphy has identified the unique characteristics of the off-base Redneck Warrior. No Redneck Warrior can possess *all* the socio-economic beliefs, habits, and traits enumerated by Murphy. Yet, if a warrior can truthfully answer "yes" to at least 30 percent of these statements he is a ***certified Redneck Warrior.*** In the interest of clarity, these statements are grouped into eight categories:

1. Social Enigmas of Redneck Warriors
2. Home Life of Redneck Warriors
3. Sex and Redneck Warriors
4. Forbidden Fruit and Redneck Warriors
5. Education and Redneck Warriors
6. Motor Vehicles of Redneck Warriors
7. Personal Hygiene of Redneck Warriors
8. Social Graces for Redneck Warriors

– Social Enigmas of Redneck Warriors –

You've been thrown out of the zoo for heckling the monkeys.

You think the moon landings were *faked*, but you think TV wrestling shows are *real*.

At the carnival you had your picture taken with a "Freak of Nature," but your friends can't tell which is which.

You've been thrown out of the zoo for heckling the monkeys.

You're amazed that gas stations keep their restrooms so clean.

You think people "out of your league" bowl on a different night.

You need only a few more holes punched in your card to get a freebie at the local Tattoo Parlor.

In your church the organist avoids the high notes so the coon dogs on the floor won't start howling.

You always see "ammo" on your mother's Christmas wish-list.

In your church the treasurer refuses to buy a chandelier until someone claims they can play one.

You routinely "make change" in the church offering plate.

You wonder why the stock market doesn't have a fence around it.

In your church the organist avoids the high notes so the coon dogs on the floor won't start howling.

You always see "ammo" on your mother's Christmas wish-list.

You have no need to buy a program at stock car races.

Your lifetime ambition is to own a fireworks stand.

You named all your male children after Confederate Army generals.

When giving someone directions to your home, you always include the phrase: "After you turn off of the paved road . . . ."

You think "going on a cruise" with your wife means driving her around the parking lot at McDonalds.

Someone in your family is named Rufus or Cletus.

You've left a six-pack of beer in the casket with a deceased friend.

You usually return from the trash dump with more than you left with.

Two or more of your close friends have died after shouting: "Hey, guys, watch this!"

You have two or more relatives named Bubba or Junior.

Someone in your family is named Rufus or Cletus.

– Home Life of Redneck Warriors –

Originally, your coffee table was a CATV cable reel.

You let your 16-year-old daughter smoke at the dinner table – in front of her children.

Your Halloween pumpkin has more teeth than your wife.

You let your 16-year-old daughter smoke at the dinner table – in front of her children.

Your spouse and your daughters keep their *spit cups* on the dining room table.

There is a stuffed possum *somewhere* in your home.

You refuse to watch the Emmy Awards because *Hee-Haw* never got the critical acclaim it deserved.

You've recorded 50 or more episodes of *The Dukes of Hazzard*.

Your functional TV sits on top of your non-functional TV.

Two or more of your coon dogs perished in the rubble when your front or back porch collapsed.

A guest once lit a match in your living room, and your whole home exploded right off of its wheels.

You consider possum to be "the other white meat."

There is a stuffed possum *somewhere* in your home.

Two or more of your children were delivered on a pool table.

You think a "loaded dishwasher" means your wife is drunk.

Your wife keeps the stuff from Graceland on the coffee table.

You think a "loaded dishwasher" means your wife is drunk.

Your wife's dress shoes are four ply non-skid lace-up Red Wings.

All your best salad bowls have *Miracle Whip* printed on them.

You and your coon dogs have the same favorite front yard tree.

Two or more of your coon dogs perished in the rubble when your front or back porch collapsed.

You've been in a marital custody battle over a coon dog.

Your coon dogs and your wallet are all on chains.

You think your wife's Thanksgiving dinner is the perfect time for your "pull my finger" trick.

While mowing grass in your back yard, you've found a lost truck.

You believe "taking out the trash" means taking your mother-in-law to watch Friday Night Wrestling.

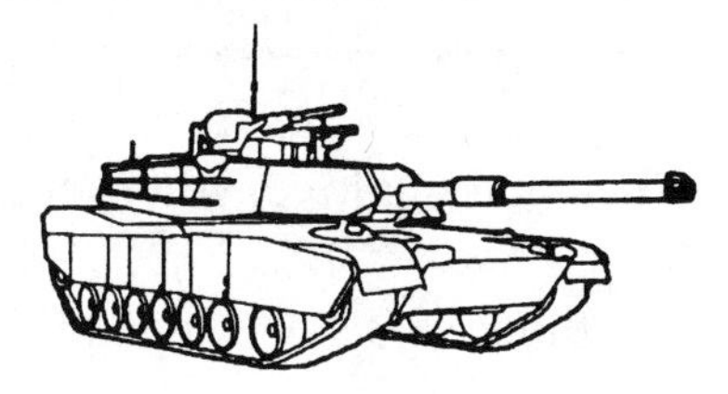

Your children took your siphon hose to school for Show & Tell.

You have a framed diploma with a filled-in blank space, followed by the words: "Truck Driving School."

While mowing grass in the back yard, you've found a lost truck.

Smith & Wesson and Redman send you Christmas cards.

All your shirts have logos printed on them.

You never removed the "Wide Load" sign from your home.

– Sex and Redneck Warriors –

You hooked up with your girlfriend after seeing her name and number on the restroom wall at the truck stop.

Your biggest turn-on is your girlfriend's beer belly.

You think women are turned-on by your skill at making seductive animal noises.

You think watching wrestling on TV is the *ultimate* form of foreplay.

You make sure the parking brake is set before making love.

> You think women are turned-on by your skill at making seductive animal noises.

While you make love, cigarettes are a no-no, but *smokeless* is OK.

If you engage in conversation while making love, it's usually just to say: "Don't worry, ain't no cars coming!"

When making love, you routinely economize by using lard.

You think *safe sex* is a padded dashboard.

You think *Genitalia* is the national airline of Italy.

– Forbidden Fruit and Redneck Warriors –

If you married your live-in girlfriend, you'd go to jail.

Your brother-in-law is also your favorite uncle.

You consider family reunions to be the ideal place to pick up some hot-to-trot women – and maybe even meet Miss Right.

You lurk under the mistletoe at Christmas and wait for your sister or your hot-to-trot Aunt Susie-Mae to wander by.

You remarry, but you still have the same in-laws.

Your father's *Last Will & Testament* leaves all his assets to his girlfriend, but she can't get the money until she turns 16.

> You consider family reunions to be the ideal place to pick up hot women, and maybe even meet Miss Right.

In your church, the congregation has four or fewer surnames.

With his trusty M-16, a modern-day warrior cranks off rounds at the bad guys, a skill easily and eagerly acquired by all Redneck Warriors (photo courtesy of U.S. Army).

Two or more women in your family have claimed their pregnancy was the result of an alien abduction.

You have a gene pool – *sort of* – but it has no deep end.

You have a five-generation family tree – with no branches.

You believe incest is OK if you keep it in the family.

– Education and Redneck Warriors –

While in the Seventh Grade you dated your father's current wife.

Fourth Grade was the best three years of your life.

While in the Seventh Grade you dated your father's current wife.

You claim the Eighth Grade in Junior High as your senior year.

Your school set up a day-care center for your Junior High prom.

Your Junior High theme song was "Friends in Low Places."

You're the local authority on ways to sneak liquor into Junior High sporting events.

Fourth Grade was the best three years of your life.

You stare at the grape-juice bottle because it says: "Concentrate."

– Motor Vehicles of Redneck Warriors –

You've *totaled* over 75 percent of the vehicles you've owned.

At least five times you've been on TV to describe your miraculous survival in yet another car wreck.

Your *home* is mobile, but most of your *vehicles* are not.

You have at least three vehicles up-on-blocks in your front yard.

Your *home* is mobile, but most of your *vehicles* are not.

Your home doesn't have shades or curtains, but your truck does.

You know exactly how many bales of hay your wife's car will hold.

Your truck's Blue Book value depends upon how much gas is in it.

At best, your truck is worth one-tenth the value of your coon dogs that ride in it.

The tailgate on your truck is held closed with bungee cords.

Your truck has a two-tone paint job – primer red and primer gray.

You've repainted your truck with a brush and house paint.

The rear tires on your truck are twice as wide as the front tires.

Your truck has a shop rag where the gas cap should be.

Your home doesn't have shades or curtains, but your truck does.

You treasure your personalized license plate because it was made by your father when he was "gone for a few years."

You never need to pull off at a Rest Stop as long as you keep those empty milk jugs handy.

You think the final line of the National Anthem is: "Gentleman, start your engines!"

You never need to pull off at a Rest Stop as long as you keep those empty milk jugs handy.

– Personal Hygiene of Redneck Warriors –

You include "chiggers" on your list of health concerns.

Your old clothes are always rejected by the Salvation Army.

You always keep several spray-cans of *Raid* on the dining room table.

For an hour at home, you can entertain yourself with a fly-swatter.

Your wife's toilet bowl brush doubles as your back-scratcher.

You use flea-and-tick soap during each end-of-the-month bath.

However, you always have to skip the end-of-December bath, because the tub will be full of beer and ice for the party.

You keep two spray-cans of *Raid* on the dining room table.

You need (1) shoes and a (2) flashlight to go to the restroom at night.

Your coon dogs gag each time they watch you eat.

Because of your *appearance* you've been fired from your job on the garbage truck.

You pull up your jeans and tighten your belt when you see posters that read: "Say No to Crack!"

Your wife's toilet bowl brush doubles as your back-scratcher.

You include "chiggers" on your list of health concerns.

You think a *bubble bath* is the result of eating too many beans.

Redneck Warriors – who truthfully answered "yes" to at least 30 percent of the statements above – have now been positively identified. Murphy cautions each Redneck Warrior that his values and culture may not be fully understood by some of his Regular Warrior brothers-in-arms. This may become a source of social friction and may adversely affect interpersonal relationships.

From time to time in social settings, Redneck Warriors may be considered uncouth. Murphy recommends that Redneck Warriors make a good faith attempt to be considerate when socializing with Regular Warriors and their families. If Redneck Warriors follow Murphy's easy-to-understand ***Social Graces for Redneck Warriors***, below, they will minimize the chance of grossing-out their Regular Warrior brothers-in-arms:

– Social Graces for Redneck Warriors –

Your coon dogs should never eat from the table until after all guests have finished their meal.

When one of your coon dogs falls in love with a guest's leg, courtesy demands that you give them a little privacy.

When expecting overnight guests, your coon dogs should vacate the guest bedroom for the evening.

After your coon dogs vacate the guest bedroom, vacuum the sheets *before* your guests arrive.

When expecting overnight guests, your coon dogs should vacate the guest bedroom for the evening.

The centerpiece on your dining room table should not be the work product of a taxidermist.

Always say "I'm sorry" after throwing up in a friend's vehicle.

Chitlins should not be the main course for dinner guests.

Don't urinate from a moving truck that is carrying guests – especially if you are driving.

Dim your headlights for oncoming vehicles, even when a round is chambered and the deer is just standing there.

Always say "I'm sorry" after throwing up in a guest's vehicle.

It's uncouth to *lay rubber* when driving in a funeral procession.

Shoes and socks should be worn to all weddings, formal or informal.

Coon dogs and livestock are practical – but not as wedding gifts.

Rat-traps are practical, but they are *uncouth* wedding gifts.

If you're the groom, it's unwise to bring a date to your wedding.

If she's the bride, a woman should never wear a tube-top or hair rollers to her formal church wedding.

Before wedding photographs are taken, remove the toothpick from your mouth.

Before wedding photographs are taken, remove the toothpick from your mouth.

When dancing at a wedding reception, regardless of the temperature, refrain from removing your partner's undergarments.

When with female guests, never fish coins out of public urinals.

When dancing at a wedding reception, regardless of the temperature, refrain from removing your partner's undergarments.

Leave beer coolers closed during all church services.

Minor children under 16-years-of-age should abstain from (1) use of tobacco products, and (2) consumption of alcoholic beverages, while inside the church, and while on church property.

When you send your wife down the road with an empty gas can, it's unwise to push your luck by telling her to bring back beer too.

Generally speaking, it's unwise to ask the State Trooper to hold your beer while you search for your driver's license.

In many cultures, women perform all the back-breaking manual labor while men sit in the shade and drink beer – a Redneck Warrior's dream! (photo by the author, Marion Sturkey).

Most of the time it's counterproductive to drink a beer during a job interview.

Go to all public restrooms in town and scratch your sister's name out of handwritten messages that begin: "For a good time, call . . . ."

– Murphy's Dinner Etiquette Tip –
for
Redneck Warriors

When dining with a female guest, never place your spit cup adjacent to her drink cup.

(seriously, ***trust Murphy*** on this one)

Murphy's Adages for Military Computers

In the prehistoric era, warriors "computed" on their fingers. If a warrior saw three bad guys sneaking toward his cave he would hold up *three fingers* to alert and inform his cave-brothers-in-arms. This simple system worked to perfection – unless the number of bad guys exceeded the warriors's number of fingers.

In the prehistoric era, warriors "computed" on their fingers.

As the centuries rolled past, math for combat got more complex. First, warriors learned to communicate numerically in writing. Next, as warfare evolved, medieval combat engineers performed complex numerical calculations when designing ***weapons of mass destruction***: siege engines, catapults, rams, and the fortified rolling towers of the day. Their feared trebuchet could hurl a 300 pound stone over 600 feet to batter down enemy castle walls.

Military whizzes eventually developed the world's first computer, the abacus. With beads strung on two wires, five on top and two on the bottom, the abacus proved to be an invaluable aid for engineers and warriors alike. Hundreds of years later in England, William Oughtred invented the slide rule in 1622. This handy new gadget allowed complex computations to be performed more rapidly.

Military whizzes eventually developed the world's first computer, the abacus.

William Thompson, an English scientist, developed a written description of a basic analog computer in 1876. Of course the technology of the times prevented anyone from actually building one. But in 1930 at the Massachusetts Institute of Technology, Vannevar Bush used Thompson's specifications and built what he called the ***Automatic Analyzer*** – the world's first *analog* computer. This gee-whiz electronic marvel became the brain of military automated fire-control weapon systems.

[He] built what he called the ***Automatic Analyzer*** – the world's first *analog* computer.

Seven years later at Harvard University, Howard Aiken combined 78 mechanical adding machines with hundreds of electrical relays. He controlled this monstrosity with perforated player-piano type paper tape. Presto! He created a *digital* computer and named it the ***Automatic Sequence Controlled Calculator***. The U.S. Armed Forces used this new marvel during World War II for complex mathematical calculations.

The so-called ***Electronic Numerical Integrator and Computer*** was born after the war. Using over 18,000 new vacuum tubes instead of the old mechanical relays, the new computer could do 5,000 additions per second. Military computers were here to stay!

He created a *digital* computer and named it the ***Automatic Sequence Controlled Calculator***.

Today the vacuum tubes and relays have gone the way of the dinosaurs. New technology has replaced them with microchips. Computers have gotten smaller and smaller, faster and faster. They permeate all aspects of modern military life. Warriors who have a desk likely have a computer sitting on it, and all warriors have computers built into their smart weapons. The wise warrior will heed ***Murphy's Adages for Military Computers***:

To err is human, but to *really* foul things up requires a computer.

Modern military logic: if it's not in the computer, it doesn't exist.

Of course computers are *evil*, but they're a *necessary evil*.

No warrior *really* learns to curse until he uses a computer.

A modern-day "wizard warrior" who specializes in tube-launched, optically-tracked, wire-guided missiles uses a bank of laptop computers (photo courtesy of U.S. Marine Corps).

To err is human, but to *really* foul things up requires a computer.

The only language common to military programmers is profanity.

Hint for warriors: never program and drink beer at the same time.

Some people say computers are simple. They are the same people who – when asked what time it is – explain how to build a clock.

If we build houses the way military programmers write programs, the first woodpecker to come along will destroy civilization.

Those who ask programmers to write in simple English will discover programmers *don't know how* to write in simple English.

It's always easy to (1) walk on water and (2) write programs to military specifications – if both the water and specifications are frozen.

In any program, any bug that *can* creep in, *will* creep in.

Computers don't make errors – they create mischief *on purpose*.

Military computers can solve all sorts of military problems, except those things that just don't add up.

When two warriors share a computer, their software preferences will differ in every possible way.

Computer wizards never die – they just lose their memory.

Hidden somewhere, there is *always* at least one more bug.

Any system that relies on computer reliability is unreliable.

In today's commercial market, the only people guaranteed to make money are those who sell computer paper.

Bug-free software *isn't.*

The most subtle bugs cause the greatest damage.

In any program, any bug that *can* creep in, *will* creep in.

When a "debugged" program crashes, it's always because of an un-debugged bug.

All military software programs contain five or more bugs.

And, hidden somewhere, there's *always* at least one more bug.

The most frightening virus is the virus you don't know you have.

If a program doesn't work it must be (1) saved and (2) backed-up.

After a flash drive has been dropped, the likelihood that it will be stepped upon is ***directly*** proportional to its value.

Most programs, when running, are obsolete.

All new computer *systems* generate all new computer *problems*.

A computer can make as many mistakes and cause as much damage in two nanoseconds as 2,000 men who toil for 2,000 years.

If a computer (1) cable and (2) connector are picked at random, the probability they will be compatible is equal to zero.

The most frightening virus is the virus you don't know you have.

Any two programs chosen at random will be incompatible.

The faster a military computer is, the sooner it will crash.

Sooner or later, all flash drives will be dropped.

After a flash drive has been dropped, the likelihood it will be stepped upon is ***directly*** proportional to its value.

The chance a flash drive will be corrupted is ***directly*** proportional to the significance of data stored on it.

Tech support can provide you with everything *except* genuine help.

The chance of a peripheral being compatible with your computer is ***inversely*** proportional to your need for that peripheral.

The chance of software being neurotic (meaning, it develops *bugs*) is ***directly*** proportional to the confusion the neurosis will cause.

The warrior who *laughs last* obviously has a back-up flash drive.

If there are ***n*** bytes in the crucial software program, the available storage space will be ***n-1*** bytes.

Tutorials are incomprehensible to those being tutored.

Military consumer expectations will outpace advances in military software technology.

New software will never work with old hardware.

Tech support can provide you with everything *except* genuine help.

General Fault Errors can be fixed – but not by you.

Tutorials are incomprehensible to those being tutored.

Programs will expand to fill all available memory – and then some.

When you finally buy enough memory, you run out of disk space.

The number of undetectable errors is infinite.

After saving the crucial file, you'll forget *where* you saved it.

General Fault Errors can be fixed – but not by you.

The only program that runs perfectly every time will be corrupted.

Illegal Error messages only occur when your file isn't saved.

Upgrades fail as soon as the old version has been deleted.

Failure is not an option – it's included with all new software.

The only program that runs perfectly every time will be corrupted.

When you hear (1) *grinding metal* and see (2) *blue smoke* coming from your hard drive, it's too late to save your file.

When all else has failed, unplug the stupid thing!

If a program hasn't yet crashed, it's waiting for the critical moment.

New computers rarely work properly – especially when nothing is wrong with them.

Circuits protected by a fuse will blow to protect the fuse.

When all else has failed, unplug the stupid thing!

– "Famous Last Words" – about Computers

Everything that can be invented, has been invented.
[Charles H. Duell, commissioner, U.S. Patent Office, 1899]

I think there is a worldwide market for – maybe – five computers.
[Thomas J. Watson, chairman of IBM, 1943]

Computers in the future may weigh no more than 1.5 tons.
[*Popular Mechanics*, forecasting new technology, 1949]

Data-processing is a fad that won't last out the year.
[Editor of business books for Prentice Hall, 1957]

But, what is it good for?
[an engineer at the Advanced Computing Systems Division, IBM, commenting on development of the microchip, 1968]

There is no reason why anyone would want a computer in their home.

There's no reason why anyone would want a computer in their home.
[Ken Olson, founder of Digital Equipment Corp., 1977]

It's true that "a picture is worth a thousand words" – but it requires ten-million times as much memory.
[Murphy, 2006]

Murphy's Writing Rules for Warriors

In the medieval past, warriors were men of action, men of few words. They charged into the fray, killed all the bad guys and their evil cohorts, and then went home. No verbal or literary expertise was required, for warriors let their lethal skills do the talking.

As centuries rolled past, verbal communication gradually became necessary in warfare. Still later, written messages came into vogue. Warriors initially tried to keep correspondence simple. For example, after the Roman Legions of Julius Caesar vanquished the bad guys at Zela in 47 BC, Caesar's written dispatch to the faraway Roman Senate contained only six words: "I came, I saw, I conquered."

Caesar's written dispatch to the Roman Senate contained only six words: "I came, I saw, I conquered."

Sadly, communication for professional warriors grew more complicated over the years. Today most hi-tech weapon systems require huge technical manuals. A big attack is set in motion not with a verbal command, but with a gigantic Operational Order that contains hundreds of pages of written detail. In a barracks environment the modern warrior must wade through a huge pile of written instructions about an ever-increasing host of responsibilities and functions.

In addition to war-fighting expertise, warriors of today must hone their literary skills.

Today, warriors must hone their literary skills. Professional warriors must become ***Professional Writers***. They must adhere to accepted principles of grammar and composition. They must rely on tools such as Simon &

Professional warriors must become ***Professional Writers***.

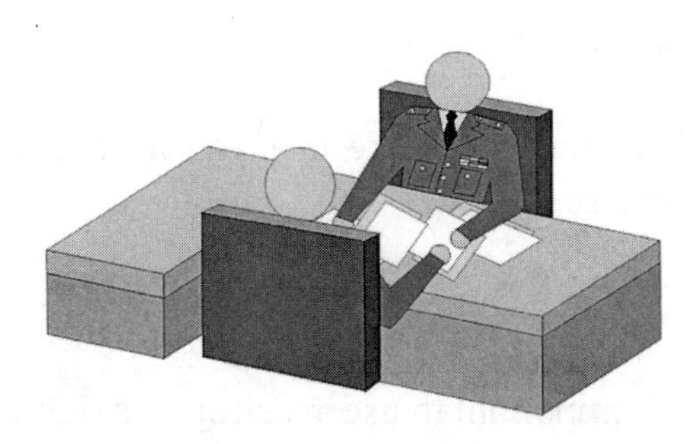

Schuster's *Handbook for Writers*, a dictionary, a thesaurus, and computer. Above all else, the wise professional warriors of today will never ignore ***Murphy's Writing Rules for Warriors***:

To avoid appearing to be ignorant, don't use no double negatives.

Remember that verbs has to agree with their subject.

Use capital letters for places like new york, atlanta, and dallas.

Badd speling an pour gramar does detract from and otherwize exsella-nt writen repport.

Each pronoun should agree with their antecedent.

Just between you and I, case is important.

Beware of sinister irregular verbs which has cropped up.

Badd speling an pour gramar does detract from and otherwize exsellant writen repport.

A good writer must not shift your point of view.

If you resort to dangling, never use participles.

If you resort to dangling, never use participles.

Always write good join clauses good like a good conjunction should.

And never use conjunctions to start your sentences.

Do not ever use a run-on sentence you have to punctuate it make sure you never forget.

A few thoughts about sentence fragments.

In letters themes reports articles and stuff like that you must use commas to keep strings of words separated.

But, do not ever, use commas, which are not, necessary.

Remember, guy's and gal's, it is important to use apostrophe's right.

Do not abbrev. in formal correspondence.

Proof your writing. Always check see if you any words out.

But, do not ever, use commas, which are not, necessary.

The author when he is writing should not get into the daily bad habit of making use of too many needless and unnecessary and extra and superfluous words which he does not really need to use in his written correspondence that he writes and that he sends to other people.

Never use a preposition to end a sentence with.

Last but not least, avoid cliches like the plague.

Murphy's Laws of Business & Finance for Warriors

Throughout recorded history the role of ***Professional Warrior*** has remained the most exalted professional calling. Warriors, the elite of their country's manhood, are revered by one and all. They proudly protect the common man, the state, and the ruling authority.

Yet, warriors know that Father Time is a stealthy killer. He gradually takes his toll. No warrior can charge into combat forever, year after year. On the modern battlefield there is scant room for senior-citizens. Advanced age soon induces warriors to trade in their cammies and combat boots for business suits and wingtips. Then they march into the amoral cutthroat civilian world of Business & Finance.

On the modern battlefield there is scant room for senior-citizens.

In the corporate world the enemy smiles and shakes hands before stabbing you in the back. The gung-ho warrior of yesteryear soon degenerates into a corporate drone. He finds that core values have changed. To prepare former warriors for this new battlefield, Murphy offers his unique insight into the world of Business & Finance.

Murphy knows that other people sometimes stumble upon a great cosmic truth, and one such person was Lawrence J. Peter. In 1969 he published an insightful book dealing with corporate inefficiency. From that book his ***Peter Principle*** is paraphrased below:

– The Peter Principle –

For every job that exists in the world there is someone, somewhere, who can not do it. Given enough time and enough promotions, he will eventually arrive at that job, and there he will remain – always bungling the job, frustrating his coworkers, and damaging the efficiency of the business.

When the warrior of yesterday enters a corporate bureaucracy he is confronted with the theoretical concept of *management*. If such a thing existed, it was misnamed. All new corporate drones should read and heed ***Murphy's Truths about Management***:

– Murphy's Truths about Management –

1. The first myth of management is that it exists.
2. The second myth of management is that skill produces success.
3. The most common managerial asset is indecision.
4. To get any decision out of management, one must create the illusion of a great crisis (such as, *the office is on fire*).
5. Managers are so wrapped-up in business *theory* that they can't comprehend business *reality*.
6. All managers claim they manage "by the book," even though they have no idea *which* book.
7. Managers whose approval is needed the most give it the least.
8. Competence varies *inversely* with the level of management.
9. Unfortunately, sales ability is assumed to be managerial ability.
10. When managers can't convince doubters, they try to confuse them.
11. The now-infamous U.S. Department of Labor study completed in May 2005 showed that 98 percent of managers predict that all current trends will continue.

When a warrior leaves the battlefield and begins his new life in a corporate cubicle, he find the rules have changed. Loyalty, honor, and moral courage are nowhere to be found. Dedication to brothers-in-arms and belief in a cause-greater-than-self are foreign concepts.

Yesterday's warrior learns new rules. He finds that the less he says, the less he has to explain. The more work he does, the more he will be given to do. Murphy sympathizes with former warriors. He offers them ***Murphy's Insight into Corporate Life***:

All things considered, corporate life is 9-to-5 against you.

Employees are usually available for work only in the past tense.

If an employee performs a difficult task efficiently (1) he will be stuck with it, and (2) he can never be promoted.

In the corporate world, stupidity is elevated to the status of a religion.

Employees are usually available for work only in the past tense.

In any meeting, progress depends upon the ratio of meeting to eating.

In any business meeting, the only surefire way to get attention is to break wind before speaking.

In the corporate world, stupidity is elevated to the status of a religion.

Infinity is a corporate lawyer waiting on another corporate lawyer.

Consultant is anyone from out of town with a briefcase.

Expert is anyone who's lucked-out on two or more wild guesses.

Investor is someone who explains that money isn't everything (for example, it isn't plentiful).

Staff is the people you need more of as incompetence increases.

Punctual employees are those who make their mistakes on time.

Adjourn is a business-meeting motion that's *always* in order.

As a former warrior settles into the routine of corporate life he learns a dreaded new word – ***committee***. In the history of the world no monument was ever erected to honor a committee. In today's corporate culture a committee may be accurately defined as:

1. The world's most efficient tool for killing time.
2. An event at which *minutes* are kept and *hours* are lost.
3. A parasitic entity with six or more legs and no brain.

– Murphy's Rules – for Corporate Committee Success

1. Never arrive on time (only amateurs are punctual).
2. Never say anything until the meeting is almost over (this makes you appear wise).
3. When you speak, be vague (this prevents irritating others).
4. Propose creation of a sub-committee (this displays your initiative).
5. Move to adjourn (this makes you popular).

The worst thing in the world is getting assigned to a committee. The only thing worse than dreaming you're in a committee meeting – and waking up to find ***you really are*** in a committee meeting – is being stuck in a committee meeting in which you can't fall asleep. Throughout history no issue of consequence has ever been resolved by a committee. The only tangible result of committee meetings is the liberation of vast amounts of hot air. A committee is a worthless cul-de-sac down which ideas are lured – and then strangled.

Murphy is privy to content of the U.S. Department of Labor study conducted in May 2005. In part, the study revealed that in 86 percent of all committee meetings, the only agreement was consensus on the next meeting date. In committee meetings the attendees exchange pleasantries, but they never think or create. Murphy documented the following ***Stages of all Committee Projects***:

– Stages of all Committee Projects –

Stage 1: *Great expectations* as the new project begins.

Stage 2: *Total disenchantment* as the new project fails.

Stage 3: *Mass confusion* as the failure becomes obvious.

Stage 4: *Frantic search* for scapegoats who can be blamed.

Stage 5: *Punishment of the innocent* to protect the guilty.

Stage 6: *Rapid promotions* for all who were not involved.

In a committee, *genuine* drivel replaces *run-of-the-mill* drivel.

In a committee, a little inaccuracy saves a world of explanation.

In a committee meeting, anyone thinking deep thoughts is thinking about lunch.

In a committee meeting, the man who smiles has thought of someone to blame.

In a committee, if a problem prompts more than one meeting, the meetings soon will become more of a problem than the problem.

Anyone thinking deep thoughts is thinking about lunch.

In a committee, if the issue is trivial and everyone understands it, the mindless chatter will last forever.

The duration of all committee meetings will be ***inversely*** proportional to the complexity of the problems discussed.

To totally stifle any novel idea, assign it to a committee.

A corporation is a giant ***bureaucracy*** staffed by drones. A bureaucracy that employs 100 or more people is a self-perpetuating empire creating so much internal paperwork that it no longer needs contact with the outside world. No person knows or cares what anyone else is doing. A former warrior finds that job security is achieved by spending all his time doing nothing – except lying about the nothing he's doing. Here and there in a bureaucracy *some* actual work is being done – by employees who have not yet risen to their level of incompetence. For the uneducated, Murphy outlines the ***Prime Characteristics of a Bureaucracy***:

Real work – if any – filters down to the lowest hierarchal level.

Competence is often more disruptive than incompetence.

The higher the *bureaucratic* level, the higher the *confusion* level.

Skill at manipulating numbers is a talent, not evidence of Divinity.

The higher the *bureaucratic* level, the higher the *confusion* level.

In any meeting, the degree of accomplishment will be ***inversely*** proportional to the number of attendees.

The duration of any meeting will be ***inversely*** proportional to the complexity of problems discussed.

The firmness of delivery dates will be ***inversely*** proportional to the tightness of the delivery schedule.

Skill at manipulating numbers is a talent, not evidence of Divinity.

Profitability will be ***inversely*** proportional to the extravagance of the annual Stockholder's Report.

Job competence will be ***inversely*** proportional to the complexity of assigned work functions.

The effort expended to defend an error will be ***directly*** proportional to the amount of damage the error caused.

Expenses and revenues will invariably rise to meet each other, no matter which one initially may have been in excess.

If there is a more time-consuming and costly way to do something, a talented bureaucrat will find it.

The standard business rule-of-thumb requires (1) adding two months to all schedules for *unexpected* delays, and (2) adding two more months for *unexpected unexpected* delays.

If it weren't for the *last minute*, nothing ever would get done.

Murphy's "**Bill of No Rights**" for Lazy Whineybabies

The American "Founding Fathers" declared independence from England in 1776 and established the United States of America. The new Constitution soon granted rights and powers to the government. Fifteen years later in 1791 a new ***Bill of Rights*** (the first ten amendments) became law. These amendments protect freedoms of speech, press, assembly, and religion. They guarantee the right to trial by jury, the right to due process, the right to bear arms, the right to petition government, etc.

Over 200 years later the lazy whineybabies in American society have become obsessed with *rights* and *entitlements*. Each day they sit on their non-working deadbeat butts and whine for (1) more money and (2) more freebie hand-outs from government – which is funded by the honest working people.

> The lazy whineybabies in American society have become obsessed with *rights* and *entitlements*.

Murphy has a solution for this degenerate welfare mentality. He has drafted a ***Bill of No Rights*** for lazy deadbeat people. These ten new amendments to the U.S. Constitution will protect the loyal conservative working people in the United States:

– Bill of No Rights –

Preamble: We – the patriotic, loyal, honest, working people of the United States – intend to form a more perfect union, establish justice, insure domestic common sense, and secure the blessings of liberty for everyone. We have worked hard, sacrificed, paid taxes, supported our government, helped the ill and the aged, and made society function for the benefit of everyone. We – the honest working people – do ordain and establish this ***Bill of No Rights*** for the lazy whineybabies, deadbeats, misfits, and perverts who have infested our country. We hold these truths to be self-evident:

ARTICLE I: You have ***no right*** to a monthly welfare check. In the United States we working people *earn* our pay-checks, and we require you to do the same. In time of illness or misfortune we will be happy to help you, but at other times you must help yourself.

You have ***no right*** to a monthly welfare check.

ARTICLE II: You have ***no right*** to free food and free housing. Working people in the United States are the most charitable people on Earth. We gladly help the infirm and the aged, the ill and the disadvantaged. Yet, we have grown sick and tired of subsidizing millions of lazy whining deadbeat couch-potato parasites who have achieved nothing in life – exclusive of breeding more and more generations of lazy whining deadbeat couch-potato parasites.

ARTICLE III: You have ***no right*** to fancy possessions such as (1) new luxury sedans, (2) new wide-screen televisions, (3) new designer clothes, etc. We working people have labored long and hard to *earn* the means to acquire such things. They are not automatically bestowed upon the sniveling and lazy idle masses.

ARTICLE IV: You have ***no right*** to free health care. We know free health care sounds like a wonderful idea to you. Yet, we working people pay for our own health care. We also pay our own housing and transportation costs, state and federal taxes, insurance and education expenses, and we support dozens

Do not complain when your next "free housing" address is the state penitentiary.

of charities. Plus, we pay the entire cost of 13 public assistance programs for you lazy and worthless deadbeat whineybabies. Therefore, there is no money left to pay for your health care too.

ARTICLE V: You have ***no right*** to happiness. In the United States we working people have only the right to *pursue* happiness, and that goes for you too. It would help if you realized that (1) happiness and (2) non-stop complaining do not go hand-in-hand.

Do not expect sympathy from us when you snap, crackle, and pop in the electric chair.

ARTICLE VI: You have ***no right*** to the possessions of others. If you rob, steal, cheat, extort, plunder, pillage, and unlawfully take away the property of honest working people, do not complain when your next "free housing" address is the state penitentiary.

ARTICLE VII: You have ***no right*** to harm others. If you attack, assault, molest, rape, abduct, kidnap, injure, mutilate, maim, or murder an innocent working person, do not expect sympathy from us when you snap, crackle, and pop in the electric chair.

ARTICLE VIII: You have ***no right*** not to be offended. Our Constitution extends freedom of speech to everyone, not just to you. If you do not like what you hear you have the right to (1) leave the room, (2) change the channel, (3) civilly express your own opinion, or (4) whine to Oprah or Jerry Springer. Otherwise, learn to live with it, and stop complaining.

You have ***no right*** to jabber in some foreign language.

ARTICLE IX: You have ***no right*** to be free from harm. When you accidentally jam a screwdriver into your eye, we working people realize that stupidity – not the screwdriver – caused the problem. Consequently, (1) try to be more careful the next time, and (2) do not expect the screwdriver manufacturer and his employees to make you a millionaire.

ARTICLE X: You have ***no right*** to jabber in some foreign language. Here, regardless of where our ancestors came from, we working people are patriotic Americans and proud of it. We are no longer Europeans, Asians, Africans, Australians, or whatever. We do not recognize any type of hyphenated-Americanism or multi-cultural multi-lingual claptrap. All proud Americans speak English *(no "Tower of Babel" problems)*. Therefore, we expect you to do three simple things: (1) ***speak English***, (2) call yourself an ***American***, and (3) assimilate into our society. Otherwise, you should return to the country from which you fled.

– Murphy's Favorite Statement –
about the
English Language
and
American Patriotism

There can be no divided allegiance. This is a nation, not a polyglot boarding house. There is no room for a fifty-fifty American, no room for a hyphenated American! *A hyphenated American is not an American at all*. We have room for one flag, the American flag. *We have room for one language, the English language*.

[Theodore Roosevelt, 22nd President of the United States]

Murphy's Justice: Susan vs Santiago and Emilio

Santiago, age 32, and his partner-in-crime ***Emilio***, age 26, were illegal aliens. In other words, these criminals had entered the United States in violation of federal law. Further, once inside this country they had embarked on a spree of violent crime. In July 2008 they apparently had watched a large secluded two-story home off Highway 84 south of Lubbock, Texas, for several days. A huge swimming pool, professional landscaping, and a big Lincoln Town Car in the driveway indicated the house would be full of valuables they could convert into cash.

> Once inside this country they had embarked on a spree of violent crime.

Tuesday, 22 July 2008: Early in the morning the homeowner and his wife left to attend a meeting at nearby Texas Tech University. The only person still in the home was their young daughter, Susan, a recent eighth grade graduate.

Santiago and Emilio probably thought Susan would be helpful. Before they killed her, they could force her to tell where all valuables in the home were located.

The two criminals had no way to know that Susan – 14 years of age and weighing 92 pounds – was no wimp. She and her parents were avid skeet shooters. Susan had become the Texas Skeet Championship runner-up in the girls junior division. She also had taken Ti Quan Do and kick-boxing lessons for two years. Now, on summer vacation and still wearing her

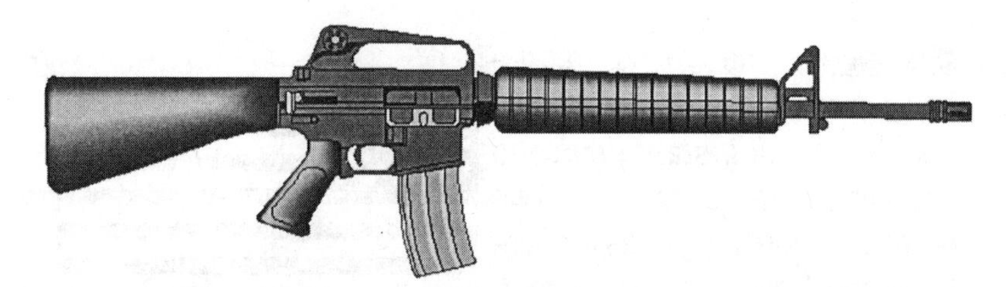

A standard M-16 assault rifle, one of Murphy's tools of choice for basic home defense purposes (illustration courtesy of U.S. Marine Corps).

pajamas, she was watching TV in her upstairs bedroom when the bad guys kicked-down the front door.

He tumbled face-down onto the floor at the bottom of the stairs, twitched a few times, and then died.

Susan leapt out of bed, dashed to her parent's room, and grabbed her dad's Remington, Model 870, pump-action 12 gauge shotgun. As Santiago burst into the room and lunged at her, she fired at point-blank range. The zero-zero buckshot tore through Santiago's lower abdomen and genitals. He screamed once, then crumpled to the floor and lay there motionless, lifeless.

Susan jumped over the body, ran into the hallway, and spotted Emilio frantically fleeing down the stairs. She racked the Remington and fired. Hit squarely in the back, Emilio tumbled face-down onto the floor at the bottom of the stairs, twitched a few times, and died.

Santiago: At the time of his death, Santiago was armed with a .45 caliber Colt, Model 1911, pistol that he had stolen during a home invasion. He had killed the 93-year-old homeowner by stabbing him four times in the chest.

Emilio: At the time of his death, Emilio was a fugitive. He had been jailed in April after the rape and mutilation of a nine-year-old child. During a bond hearing he overpowered his guards, stole their handguns, shot two police officers, and escaped.

Susan: The young freckle-faced teenage girl with a ponytail had dished-out ***instant justice*** to two criminal psychopaths. Two down! No more burglaries for Santiago and Emilio. No more murders. No more thefts. No more victims. No deportation hearings. No more escapes. No more trials. No appeals. No more incarceration at taxpayer expense.

On Saturday, four days after she exterminated the two criminals, Susan and her parents held a news conference. Susan thanked her mom and dad for teaching her assertiveness and self-defense skills. With a modest grin, she told newspaper reporters that the two dead criminals had "tried to pick on the wrong girl."

At that point a liberal media whore stood up to ask a question. She asked why Susan had displayed absolutely "no sympathy" for the two "undocumented immigrants."

Susan's face turned red, and her eyes flashed with anger. "What an idiot!" she thought. She leaned forward and looked straight at the brain-dead media whore. With the TV cameras rolling, 14-year-old Susan offered an all-time classic response:

> Undocumented? Undocumented? Lady, those perverts weren't undocumented – and they weren't immigrants either. They were illegal aliens! They were thieves! They were murderers! They tried to kill me! If you're so stupid that you think they were *undocumented immigrants*, you probably think people who sell crack-cocaine are *undocumented pharmacists*.

Murphy's Journey: Young Warrior to Old Veteran

. . . a few serious thoughts about growing old:

Young Americans of the 1940s, 1950s, 1960s, and 1970s grew up with Butch Wax, Green Stamps, roller skate keys, and duck-and-cover drills. They grooved on music of the Andrews Sisters, The Ink Spots, or The Supremes, and they knew Ed Sullivan had the best show on the new electronic marvel called television. These budding young adults lived in a visionary exciting age when the worst thing you usually caught from a woman was a cold.

When their country called, young patriots made the transition from civilians to American Warriors. They marched off to war in remote and God-forsaken corners of the globe carrying Zippo lighters, P-38s, steel pots, flak jackets, and heat tabs. Amid the perils and misery of combat they lived with malaria, dysentery, ringworm, leeches, and trench-foot. They were always exhausted, always hungry, always scared of dying, but too proud to let it show. By the 1970s they knew Tony Bennett had left his heart in a mystical place called San Francisco. Many vowed to go there and wear flowers in their hair if they survived their war.

> They were always exhausted, always hungry, always scared of dying, but too proud to let it show.

In battle they carried the memory of friends with whom they had shared life in one instant, but who had been shot, mortared, bombed, burned, rocketed, or blasted into oblivion in the next. They carried a mere nostalgic affection for the

> They fought, lived, and often died in mud and filth with the rats and unburied corpses of their friends.

Early on the morning of 8 April 1942, aerial warriors of yesteryear get ready to take off from the USS Hornet in their B-25 Mitchell bombers for the famous and perilous "Doolittle Raid" against Tokyo, Japan (photo courtesy of U.S. Navy).

world they had left behind, but an intense eternal loyalty to their old brothers-in-arms.

The oldest warriors remember faraway evil places like Iwo Jima, Omaha Beach, and a tyrant named Hitler. They fought, lived, and often died in mud and filth with the rats and unburied corpses of their friends. A later generation stood its ground during the war in frozen Korea and during the long Cold War. When the bitter struggle in Vietnam raged, they ignored pathetic whinybabies at home and answered the call of their country during her darkest hours since her Civil War in the previous century. When megalomaniacs and terrorists enslaved millions in the Middle East, a new generation of American Warriors risked their lives – and often lost them – in an effort to restore freedom for people they would never know.

> Warriors who survived came home from their wars and held jobs, raised families, paid taxes, and made society function.

> American Warriors, not the media whores, preserved our freedom of the press.

Historically, when enemies attack our country it has been American Warriors, not media whores, who preserve our freedom of the press. It has been American Warriors, not church pastors, who preserve our freedom of religion. When despots assail our society it has been American Warriors, not lawyers, who preserve the civil liberties we enjoy. It has been American Warriors, not so-called civil rights activists, who preserve freedom of speech for all. It has been the American Warrior, not the posturing pompous politician, who often has sacrificed his life so that others might live in peace and freedom.

> The aging military veteran of today is the average older man, a neighbor, an uncle, maybe a grandfather.

The warriors who survived came home from their wars and held jobs, raised families, paid taxes, and made society function. As the years roll by these aging warriors of yesteryear find they have evolved into elderly men, old farts, old military veterans, sometimes even genuine geezers. To their surprise they discover they have gained respectability. Even teenagers admire them, for society now realizes old warriors have paid their dues in full.

To a man, old military veterans despise clamoring media whores and pitiful apostles of political correctness. They put their faith in their God, their country, their family, and their old brothers-in-arms. They rely on family values, hard work, a good education, and common sense. Freedom is not free. Liberty entails responsibilities, and old military veterans know it.

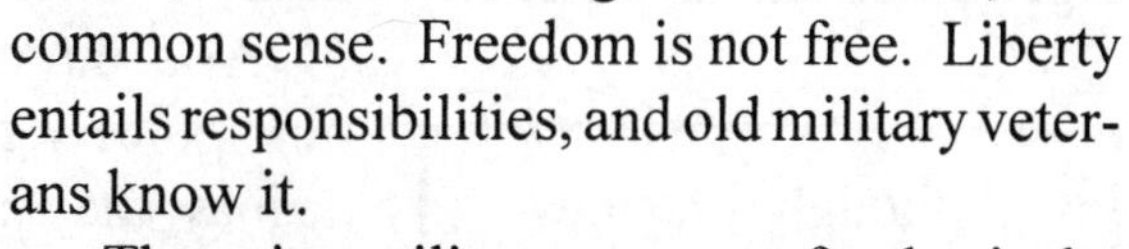

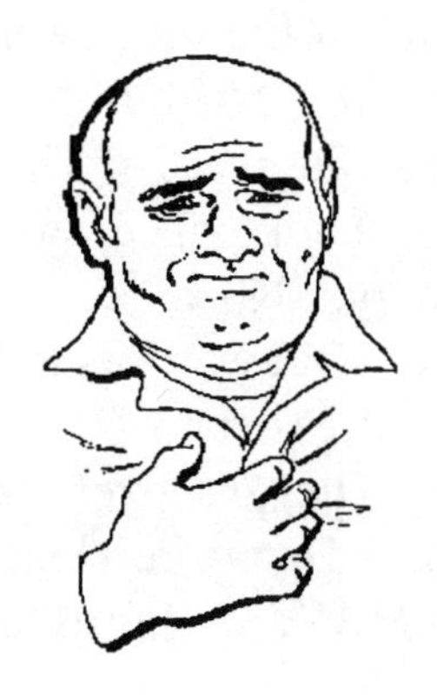

The aging military veteran of today is the average older man, a neighbor, an uncle, maybe a grandfather. He is the same man who, in his youth when his country called, gave selflessly of himself and asked for nothing in return. Today's old military veteran is the now-wrinkled warrior of days gone by. He is the man who was willing to sacrifice the flower of his youth so that people he would never know might survive and live in peace and freedom.

These old warriors open doors for women, and they don't like the filth and trash displayed on TV.

Today at sporting events when the National Anthem is played, old military veterans *sing*. They know the words, and they *believe* in them. These old warriors open doors for women, and they don't like the filth and trash displayed on TV. They have moral courage, and they brag only when conversation turns to their old brothers-in-arms or their children and grandchildren.

For American Warriors of yesteryear who have evolved into the old military veterans of today, life is good. Old military veterans get *respect*, which in most cases was long overdue. American society still needs them, their common sense, their work ethic, their family values, and their loyalty to our country – now more than ever.

. . . thoughts from Murphy about growing old:

Despite the high cost of *living*, it remains popular. Most old warriors of yesterday have adopted the premise that they intend to live forever, and so far, so good. Although growing old is mandatory, growing up is optional. Warriors age, but their definition of Middle Age or Old Age is constant – about ten years older than they are.

Despite the high cost of *living*, it remains popular.

Although time may have been a great healer, it turned out to be a lousy beautician.

At some point in life older warriors go *over the hill*. Then, to their dismay they pick up speed. They know they were born naked, wet, scared, and hungry. Now, year by year it gets worse. Old warriors find that it's been a lot easier to get old than to get wise. They discover that although time may have been a great healer, it turned out to be a lousy beautician. Murphy has scientifically categorized ***Middle Age*** as the time in your life when:

– Murphy's Criteria -- for Middle Age

You choose your cereal for the fiber, not for the toy.

You often stop to think – and forget to start again.

You still believe you'll feel better in the morning.

You decide that the best exercise is discretion.

You're too young for Social Security, but too old for a job.

You still can do just as much as ever – but you'd rather not.

You're at home alone on Friday night, your phone rings, and you hope it's not for you.

You stopped whining about the older generation years ago, and now you whine about the younger one.

Your broad mind and your narrow waist have exchanged places.

Warriors of yesteryear never worry too much about Middle Age, and they quickly outgrow it. Then they face the yawning abyss of Old Age. It's a terrible price to pay for presumed wisdom. There isn't much future in Old Age. Old warriors finally know most of life's answers, but no one is asking the questions. They discover that life is like a roll of toilet paper – the closer you get to the end, the faster it goes. They've learned ***Old Age*** is the time in life when:

You're entering the prime of senility.

You're entering the prime of senility.

You still could touch your toes – if they were on your knees.

Your *wild oats* have matured into *prunes and cereal*.

You're still better than dead – more or less.

You can smile, because you know tomorrow will be worse.

The annual cost of your candles exceeds the cost of your cake.

You approach problems with an *open mind* and an *open fly*.

You finally have your head together, but your body is falling apart.

You get absent-minded. You get absent-minded. You get . . . .

You wish "the buck stopped here" because you need the money.

It's easy to meet expenses – they're everywhere.

You plan to be spontaneous and carefree – tomorrow.

It's hard to make a comeback – because you haven't been anywhere lately.

The world beats a path to your door only if you're in the bathroom.

You spend more time wondering about the "hereafter." You go into the kitchen or bedroom to get something – and then wonder what you're "hereafter."

Your new *clear conscience* is the result of your new *faulty memory.*

You want to know that, if all is not lost – where is it?

Your new *clear conscience* is the result of your new *faulty memory.*

You plan to be spontaneous and carefree – tomorrow.

In your youth, you ***had long hair***. Now, you ***long for hair***.

In your youth, passing the ***driving test*** was a big deal. Now, passing the ***vision test*** is a big deal.

In your youth, ***acid rock*** was the thing. Now, ***acid reflux*** is the thing.

You often have to change your underwear after each sneeze.

In your youth, you were obsessed with the ***Rolling Stones***. Now, you are obsessed with your ***Kidney Stones***.

In your youth, you wanted to move to California because it was ***cool***. Now, you want to move to California because it's ***warm***.

In your youth, you experienced the fun of going to a ***new hip joint***. Now, you experience the agony caused by your ***old hip joint***.

It can no longer be denied. Old Age has set in. The carefree young warrior of yesterday finds he has evolved into an ignominious icon of today, a gen-u-ine Geezer. Being a geezer has drawbacks. You often have to change your underwear after each sneeze. And if a good-looking woman suggests, "Let's go upstairs and make love," you are forced to reply, "Pick *one*, I can't do both!"

Although being "over the hill" may be bad, it's still better than being *under* it.

Yet, there are plenty of perks for geezers. Life is not totally the pits. Although being "over the hill" may be bad, it's still better than being *under* it. Your supply of brain cells is down to a manageable level, and your eyes can't possibly get much worse. You enjoy hearing about your friends' operations, and your only heated arguments are about Social Security. A ringing telephone never bothers you after 9:00 pm because nobody you know is awake that late.

Actually, geezers are lucky in many ways. Being a geezer isn't all doom and gloom. Below, Murphy has listed some of the benefits of achieving geezer status:

– Murphy's Benefits – for Geezers

The things you buy now won't ever wear out.

You have nothing left to learn the hard way.

Your investment in health insurance is finally starting to pay off.

Kidnappers are no longer interested in you.

In a mass hostage situation, you'll be released first.

You no longer have any reason to try to hold your stomach in.

For a woman, going bra-less pulls the wrinkles out of her face.

Your joints are better meteorologists than the TV weatherman.

Your doctor, not the police, tells you to slow down.

"Getting a little action" means you don't need more fiber.

"Getting lucky" means you found your car in the mall parking lot.

You no longer care where your spouse goes.

You finally can live without sex – but not without glasses.

A geezer's secrets are totally safe with all of his friends – they can't remember them either.

– PART THREE –

Heritage of the American Warrior

(Tribute to U.S. Warriors)

Anthems of U.S. Armed Forces and National Anthem

– United States Army –

No military anthem of the United States has undergone more change than the U.S. Army anthem. In 1908 the Army had stationed 1stLt. Edmund L. Gruber with the 5th Field Artillery in the Philippines. In that bygone era the field artillery caissons were pulled by teams of mules. While accompanying an artillery detachment on primitive roads through rugged mountains, Gruber heard the section chief repeatedly yell to his artillerymen: "Keep 'em rolling!"

Gruber and two friends, William Bryden and Robert Danford, later wrote an inspirational song, and Gruber composed a peppy tune. Words of their upbeat anthem, ***The Caisson Song***, relate to the mule-drawn artillery caissons "rolling along." The song became wildly popular with artillerymen in the Philippines. Soldiers returning to the United States brought the song home with them, and soon all six regiments in the U.S. Army Field Artillery unofficially adopted it:

> In that bygone era the field artillery caissons were pulled by teams of mules.

– The Caisson Song –

(original "caisson" lyrics)

Over hill, over dale,
As we hit the dusty trail,
And those caissons go rolling along.
In and out, hear them shout,
Counter march and right about,
And those caissons go rolling along. *[Sing Refrain]*

[Refrain]
Then it's hi! hi! hee!
In the field artillery,
Shout out your numbers loud and strong,
For where e'er you go,
You will always know,
That those caissons go rolling along.

In the storm, in the night,
Action left or action right,
See those caissons go rolling along;
Limber front, limber rear,
Prepare to mount your cannoneer,
And those caissons go rolling along. *[Sing Refrain]*

Then it's hi! hi! hee!
In the field artillery.

Was it high, was it low,
Where the hell did that one go?
As those caissons go rolling along;
Was it left, was it right?
Now we won't get home tonight,
And those caissons go rolling along. *[Sing Refrain]*

During World War I the Army hierarchy wanted a marching song. Noted bandmaster John Philip Sousa, believing *The Caisson Song* dated back to the Civil War era, made minor changes to the words and tune and retitled it, *The U.S. Field Artillery March.* The song became popular with the Army and the American public and sold almost a million copies. Sousa soon learned that Gruber had written the song only a few years before, so he did the proper thing. He arranged for Gruber to get all credit and monetary royalties.

In 1948 the Army conducted a contest to find an updated Army song.

In 1948 the Army conducted a contest to find an updated Army song – no more mule-drawn caissons. The submissions were dismal, so no selection was made. Four years later the Army asked the civilian music industry for help. Over 700 compositions poured in, but none had the *magic* the Army sought. Finally an American soldier, H.W. Arberg, hit paydirt. He kept the tune of *The Caisson*

Song, but he wrote new lyrics to bring the ditty up to date. The Army brass knew a musical winner when they saw one and immediately adopted the new song. The new anthem, ***The Army Goes Rolling Along***, is now the official song of the U.S. Army:

– The Army Goes Rolling Along –

(new "first to fight" lyrics)

[Introduction]
March along, sing our song, with the Army of the free,
Count the brave, count the true, who have fought to victory,
We're the Army and proud of our name!
We're the Army and proudly proclaim:

First to fight for the right,
And to build the Nation's might,
And the Army goes rolling along.
Proud of all we have done,
Fighting 'till the battle's won,
And the Army goes rolling along.

[Refrain]
Then it's Hi! Hi! Hey! The Army's on its way!
Count off the cadence loud and strong,
For where e'er we go, you will always know,
That the Army goes rolling along.

Valley Forge, Custer's ranks,
San Juan Hill and Patton's tanks,
And the Army went rolling along,
Minute-men, from the start,
Always fighting from the heart,
And the Army goes rolling along. *[Sing Refrain]*

The Army goes rolling along.

Men in rags, men who froze,
Still that Army met its foes,
And the Army went rolling along,
Faith in God, then we're right,
And we'll fight with all our might,
As the Army goes rolling along. *[Sing Refrain]*

– United States Navy –

The U.S. Navy anthem also changed. Lt. Charles Zimmermann had become bandmaster of the U.S. Naval Academy Band in 1887. He began composing a musical march each year in honor of that year's Naval Academy graduating class.

In 1906, Alfred H. Miles, an Academy midshipman, suggested a special "football song" for the upcoming Army-Navy football game. Miles thought their underdog team needed an inspirational song that would "live forever." It seemed like a good idea to Zimmermann, so he composed a jaunty tune. Collaborating with Zimmermann, Miles wrote lyrics for two stanzas of ***Anchors Aweigh***, a new football team "fight song":

An Academy midshipman suggested a special "football song" for the upcoming Army-Navy football game.

– Anchors Aweigh –

(original football lyrics)

Stand, Navy, down the field, sails set to the sky,
We'll never change our course, so, Army, you steer shy-y-y-y,
Roll up the score, Navy, Anchors Aweigh,
Sail, Navy, down the field, and sink the Army,
sink the Army Grey.

Get underway, Navy, decks cleared for the fray,
We'll hoist true Navy Blue, so, Army, down your Grey-y-y-y,
Full speed ahead, Navy; Army, heave to,
Furl Black and Grey and Gold, and hoist the Navy,
hoist the Navy Blue.

At the annual Army-Navy football game in November 1906 the Naval Academy Band played *Anchors Aweigh*, and Navy midshipmen sang at the top-of-their-lungs. For the first time in several years, Navy won the game. To celebrate their victory the Naval Academy dedicated the song to the entire Class of 1907. Later the U.S. Navy

The famed USS Constitution, commissioned on 21 October 1797, is the world's oldest commissioned warship still afloat. Here, the ship sails unassisted in Massachusetts Bay on 21 July 1997 (photo courtesy of U.S. Navy).

adopted *Anchors Aweigh* as the official Navy anthem.

The original song, written for the football game, contained lyrics intended to motivate the team. The words didn't have much to do with military responsibilities and duties of the Navy. To correct this situation, George Lottman later penned replacement lyrics for the two original stanzas. The new lyrics made *Anchors Aweigh* an ideal nautical anthem for the Navy. Royal Lovell, another midshipman, wrote the final stanza in 1926. The second stanza of the revised lyrics has evolved into a beloved musical icon of the U.S. Navy. Also, because the tune is unchanged, *Anchors Aweigh* remains the Naval Academy football team "fight song":

The new lyrics made *Anchors Aweigh* an ideal nautical anthem for the Navy.

– Anchors Aweigh –

(new military lyrics)

Stand, Navy, out to sea; fight, our battle cry,
We'll never change our course, so vicious foe, steer shy-y-y-y,
Roll out the TNT, Anchors Aweigh,
Sail on to victory, and sink their bones to Davy Jones, hooray!

Anchors Aweigh, my boys, Anchors Aweigh,
Farewell to college joys, we sail at break of day-y-y-y,
Through our last night on shore, drink to the foam,
Until we meet once more, here's wishing you a happy voyage home.

Blue of the Seven Seas; Gold of God's great sun,
Let these our colors be, 'till all of time be done-n-n-n,
By Severn shore we learn, Navy's stern call:
Faith, courage, service true, with honor over, honor over all.

– United States Marine Corps –

The oldest United States military anthem belongs to the U.S. Marine Corps, but the Marines are different. Instead of a song, they have a hymn. ***The Marines' Hymn*** chronicles military prowess of the Marine Corps. During the war with Barbary Pirates in 1805, Lt. Presley O'Bannon led a rag-tag band of Marines and mercenaries and captured the fortress at Derna on *the shores of Tripoli*. During the Mexican War, Marines led the charge to seize Chapultepec Castle, the ancient *halls of Montezuma*. Reversing these two phrases in the interest of euphony, an unidentified author penned the first line of the world's most famous military anthem.

The Marines' Hymn was in widespread use by Marines and the American public by the mid-1800s. Nobody was able to identify the person who wrote the lyrics, but various people tried to identify the origin of the tune. Col. A.S. McLemore assumed the tune came from the comic opera *Genevieve de Barbant*, which had been presented in Paris, France, on 19 November 1859. Yet,

> Instead of a song, they have a hymn.

many others claim the tune has earlier roots in a Spanish folk song.

The Marines' Hymn became the official anthem of the Marine Corps in 1929. Thirteen years later on 21 November 1942 the Marine Commandant approved a change in the words of the first stanza, fourth line. Because of increasing use of aircraft in the Corps, the words "in the air" were added to preface "on land, and sea."

The combat-and-victory theme and peppy tune of the hymn made it famous worldwide. Sir Winston Churchill (1874-1965), British Prime Minister, became fascinated with the hymn. At functions of state he often entertained guests by reciting, from memory, all three stanzas of *The Marines' Hymn*:

– The Marines' Hymn –

From the Halls of Montezuma,
To the Shores of Tripoli;
We fight our country's battles,
In the air, on land, and sea;
First to fight for right and freedom,
And to keep our honor clean;
We are proud to claim the title,
– of United States Marines.

Our flag's unfurled to every breeze,
From dawn to setting sun;
We have fought in every clime and place,
Where we could take a gun;
In the snow of far-off northern lands,
And in sunny tropic scenes;
You will find us always on the job,
– the United States Marines.

Here's health to you and to our Corps,
Which we are proud to serve;
In many a strife we've fought for life,
And never lost our nerve;
If the Army and the Navy,
Ever look on Heaven's scenes;
They will find the streets are guarded,
– by United States Marines.

– United States Air Force –

In the early 1900s, like most other countries, the United States did not have a separate Air Force. Instead, it had an "Air Corps" that was part of the U.S. Army. The U.S. Army Air Corps had no separate anthem. The unofficial anthem of the Army at that time, *The Caisson Song*, was oriented toward mule-drawn ground artillery.

In 1938, *Liberty Magazine* sponsored a contest to find an official song for the Army Air Corps. The magazine received 757 entries. The selection committee picked Robert M. Crawford's *Off We Go into the Wild Blue Yonder*. Crawford, an accomplished songwriter and vocalist as well as a civilian pilot, introduced and sang his new song at the Cleveland Air Races on 2 September 1939.

The Army Air Corps faded into history on 18 September 1947.

The Army Air Corps faded into history on 18 September 1947. By congressional mandate, the new U.S. Air Force then absorbed the men and machines of the nation's air arm. The new U.S. Air Force also adopted ***Off We Go into the Wild Blue Yonder*** and replaced the last words of each stanza, "Army Air Corps," with "U.S. Air Force."

"A Toast to the Host" is part of the Air Force song. This musical bridge, with a different melody and mood, is a tribute to warriors of the Air Force who have been killed-in-action:

– Off We Go into the Wild Blue Yonder –

Off we go into the wild blue yonder,
Climbing high into the sun;
Here they come zooming to meet our thunder,
At 'em, boys, give 'er the gun!
Down we dive spouting our flames from under,
Off with one hell-uv-a roar!
We live in fame, or go down in flame,
Nothing can stop the U.S. Air Force!

We live in fame, or go down in flame.

The musical score of Off We Go into the Wild Blue Yonder was carried to the moon on 30 July 1971 aboard the Apollo 15 lunar module. With an audiotape player, American astronauts broadcast Off We Go' to the world as they blasted off from the moon to dock with the orbiting command module (photo courtesy of NASA).

Minds of men fashioned a crate of thunder,
Sent it high into the blue,
Hands of men blasted the world asunder,
How they live, God only knew,
Souls of men dreaming of skies to conquer,
Gave us wings ever to soar!
With scouts before and bombers galore,
Nothing can stop the U.S. Air Force!

Nothing can stop the U.S. Air Force!

[**Musical Bridge**]
Here's a toast to the host of those,
Who love the vastness of the sky;
To a friend we send the message,
Of his brother-men who fly.

We drink to those who gave their all of old,
Then down we roar to score the rainbow's pot of gold.
A toast to the host of men we boast, the U.S. Air Force!

Off we go into the wild blue yonder,
Keep the wings level and true;
If you'd live to be a gray haired wonder,
Keep your nose out of the blue!
Flying men guarding our nation's borders,
We'll be there followed by more,
In echelon we carry on,
Nothing can stop the U.S. Air Force!

Flying men guarding our nation's borders.

– United States of America –

Bitter territorial squabbles and trade disputes boiled over into war between the United States and Great Britain in 1812. Fighting on home ground, the United States won early victories. However, in the summer of 1914 with Napoleon temporarily exiled on the island of Elba, Britain was able to turn its entire military juggernaut against the United States.

Britain was able to turn its entire military juggernaut against the United States.

That summer the British invaded Washington, DC. They burned and sacked the American capital and then turned their war machine toward nearby Baltimore, the coastal city guarded by Fort McHenry on the Patapsco River inlet. Adm. Alexander Cochrane, Royal Navy, planned for the big guns and mortars of his battle fleet to pound Fort McHenry into rubble. Then his infantry would debark, storm the fortifications protecting Baltimore, and seize the city.

Meanwhile an American attorney, ***Francis Scott Key***, had sailed out to the British fleet under a flag of truce.

Meanwhile an American attorney, ***Francis Scott Key***, had sailed out to the British fleet under a flag of truce. Key was able to negotiate an exchange of prisoners-of-war. Some of the American prisoners were held aboard British warships. The British agreed to release them into Key's custody after the coming battle, and they forced Key to remain overnight aboard a British warship.

Shortly after noon on 13 September the Royal Navy opened fire. Safely beyond the range of Fort McHenry's guns, British warships blasted American defenders. Sixteen lethal "bomb ships" lobbed huge 185-pound high-explosive shells toward Fort McHenry. Fired from over two miles away, these high-trajectory siege mortars rained destruction down onto the Americans.

Just before dawn the British guns and siege mortars fell silent. Had the fort surrendered? Had the American flag been hauled down?

Looking through his telescope, Key anxiously watched the bombardment. The symbol of American defiance – the huge American flag with 15 stars and 15 stripes, approved by an act of Congress in 1794 – was visible "at the twilight's last gleaming." All through the night the nonstop shelling continued. In a sky illuminated by "the rockets' red glare, the bombs bursting in air," Key periodically could see the flag atop the flagpole inside Fort McHenry. Thus far there had been no capitulation.

Just before dawn the British guns and siege mortars fell silent. Had the fort surrendered? Had the American flag been hauled down? Soon aided "by the dawn's early light," Key peered through his telescope. He saw the tattered flag of his country still defiantly fluttering in the stiff morning breeze. Elated and inspired, he began to compose a poem on the back of an envelope.

Thwarted, the British fleet sailed away, and Key headed back to Baltimore. In a hotel

The public began singing Key's spirited poem to the tune of a bar-room ballad, *To Anacreon in Heaven.*

room he edited his poem and titled it, *The Defense of Fort McHenry.* He went to a local printer who quickly prepared hundreds of copies of the poem on handbills. Key proudly posted them throughout the city of Baltimore.

To celebrate the victory the public began singing Key's poem to the tune of a bar-room ballad, *To Anacreon in Heaven.* The song's popularity grew, and a public performance took place the next month. The song gradually became known as *The Star Spangled Banner*.

The song gradually became known as *The Star Spangled Banner.*

In 1931, Congress finally adopted ***The Star Spangled Banner*** as the National Anthem of the United States. Today the tattered American flag that flew over Fort McHenry in 1814 is preserved in the Smithsonian Museum in Washington, DC:

– The Star Spangled Banner –

O, say, can you see by the dawn's early light,
What so proudly we hail'd at the twilight's last gleaming?
Whose broad stripes and bright stars, thro' the perilous fight,
O'er the ramparts we watch'd, were so gallantly streaming.
And the rockets' red glare, the bombs bursting in air,
Gave proof thro' the night that our flag was still there.
O, say, does that star-spangled banner yet wave,
O'er the land of the free and the home of the brave?

And the rockets' red glare, the bombs bursting in air, Gave proof thro' the night that our flag was still there.

On the shore, dimly seen thro' the mists of the deep,
Where the foe's haughty host in dread silence reposes,
What is that which the breeze, o'er the towering steep,
As it fitfully blows, half conceals, half discloses?
Now it catches the gleam of the morning's first beam,
In full glory reflected, now shines on the stream:
'Tis the star-spangled banner! O, long may it wave,
O'er the land of the free and the home of the brave!

And where is that band who so vauntingly swore,
That the havoc of war and the battle's confusion,
A home and a country, shall leave us no more?
Their blood has wash'd out their foul footsteps' pollution.
No refuge could save the hireling and slave,
From the terror of flight or the gloom of the grave:
And the star-spangled banner in triumph doth wave,
O'er the land of the free and the home of the brave!

O, thus be it e'er when free men shall stand,
Between their lov'd homes and the war's desolation!
Blest with vict'ry and peace, may the Heav'n-rescued land,
Praise the Pow'r that has made and preserv'd us a nation!
Then conquer we must when our cause is just,
And this be our motto: "In God is our trust!"
And the star-spangled banner in triumph shall wave,
O'er the land of the free and the home of the brave!

And the star-spangled banner in triumph shall wave,
O'er the land of the free and the home of the brave!

The History of Blood Chits

A ***Blood Chit*** is a tool that America's combat aircrews always want to have, although they hope they never have to use it. It's the common name for a written notice, printed in several languages, carried by American military aircrews in combat. If their aircraft is shot down or forced down in hostile territory, the notice (1) identifies the airmen as Americans, (2) encourages the civilian population to assist them, and (3) offers a reward for such assistance.

This concept is over 200 years old.

This concept is over 200 years old. Jean-Pierre Blanchard, the famous French balloonist, came to America in 1793 to demonstrate hot air balloon flight. He would ascend from Philadelphia. Where he would come down, of course, no one knew. Further, Blanchard did not speak English. George Washington, the U.S. President, gave Blanchard a letter addressed to "All citizens of the United States." The letter asked Americans to befriend Blanchard and help him safely return to Philadelphia.

During World War I the British Royal Flying Corps gave ***Ransom Notes*** to its combat aircrews flying in India and Mesopotamia.

Thereafter this idea lay dormant for over a century. Then, in 1917 during World War I the British Royal Flying Corps gave ***Ransom Notes*** to its combat aircrews flying in India and Mesopotamia. Ransom notes featured a notice printed in four languages: Arabic, Urdu, Farsi, and Pashto. The notice promised a reward to anyone who brought an "unharmed" British pilot or observer to the nearest British outpost. Some hostile tribesmen had been *castrating* captured airmen, so ransom notes included the word "unharmed." With a touch of gallows

The American pilots carried a printed notice called a ***Blood Chit***.

Three old Stearman biplanes, slightly more advanced than warplanes used by the Royal Flying Corps in India, fly over Fort Sam Houston in Texas (photo courtesy of U.S. Army).

humor, British fliers called the notes ***Goolie Chits***, because *goolie* was an English slang term for *testicle*.

When the American "Flying Tigers" went to China in 1937 to battle the invading Japanese Army and Air Force, the American pilots carried a printed notice called a ***Blood Chit***. These notices were sewn onto the back of pilots' flight jackets. They bore the Chinese flag and Chinese symbols which conveyed the following message:

> This foreign person has come to China to help in the war effort. Soldiers and civilians, one and all, should rescue [him], protect [him], and provide him with medical care.

Over time the U.S. Armed Forces changed wording on blood chits carried by Americans fighting in China. The new wording read:

> I am an American airman. My plane is destroyed. I can not speak your language. I am an enemy of the Japanese. Please give me food and take me to the nearest Allied military post. You will be rewarded.

When the United States *officially* entered World War II in 1941, all combat aircrews carried blood chits. Each displayed the American flag. Chits were available in over 40 languages, so a flier carried the appropriate chit for his area of operations. Each chit offered a reward to any person who assisted a downed American flier.

The greatest reward ever granted would go to Yu Song Dan. North Korean forces shot down a big American B-29 bomber on 12 July 1950, two weeks after the start of the Korean War. Members of the Yu Ho Chun family found the badly injured American crewmen. Mr. Chun discovered a blood chit in the pocket of one flier, and he gave the Americans medical aid. Then, at great personal risk he put them on a junk and sailed them 100 miles down the coast to safety. Two weeks later North Korean Army soldiers found Chun, tortured him, and then killed him. The wheels of justice ground slowly, but 43 years later in 1993 the United States paid a $100,000.00 reward to Chun's son, Yu Song Dan.

North Korean Army soldiers found Chun, tortured him, and then killed him.

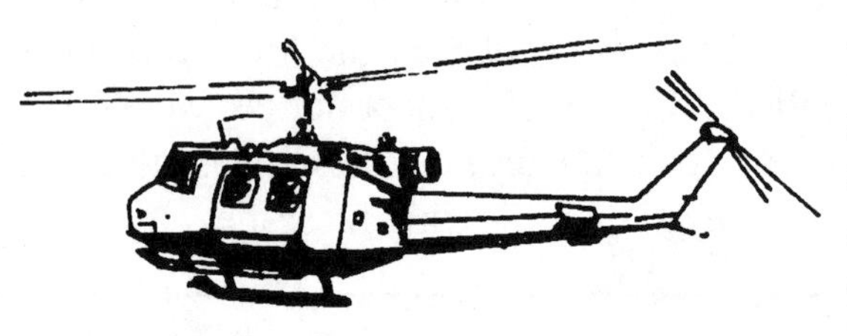

During the war in Vietnam the American fighter, ground attack, and helicopter crews carried updated blood chits. These new chits displayed the American flag plus an appeal in 14 languages: English, Burmese, Thai, Old Chinese, New Chinese, Laotian, Cambodian, Tagalog, Vietnamese, Visayan, Malayan, French, Indonesian, and Dutch. The translation in each language was the same:

> I am a citizen of the United States of America. I do not speak your language. Misfortune forces me to seek your assistance in obtaining food, shelter, and protection. Please take me to someone who will provide for my safety and see that I am returned to my people. My government will reward you.

In Vietnam and other parts of Southeast Asia, unique missions required unique measures. On clandestine cross-border flights the crewmen usually carried paper money and gold coins in addition to their blood chits. Needless to say, these required strict inventory control. Upon return from a secret *Black Ops* flight a verbal claim that the money had been "lost" would not suffice.

Today the United States stocks pre-printed blood chits for most regions of the world.

Today the United States stocks pre-printed blood chits for most regions of the world. Chits in the appropriate languages were issued to American aircrews for operations in Panama, Grenada, Somalia, and Bosnia. More recently, chits were issued for combat missions in Kuwait, Iraq, Afghanistan, and the surrounding areas. In the battle against international terrorism the current blood chit package usually includes money, gifts, and often a pointee-talkee pictorial display.

The standard Middle East blood chit bears a written notice printed in 13 languages.

For combat flights in the volatile Middle East, wording on blood chits has not materially changed since the Vietnam era. The standard Middle East blood chit bears a written notice printed in 13 languages. These include European languages such as Polish and German, and Middle East languages such as Arabic and Serbo-Croatian. The American flag and an Operation Iraqi Freedom logo are printed on Middle East chits. The United States stands by the blood chit program. As recently as 20 July 2006 an updated U.S. Department of Defense written policy reaffirmed:

> Through the Blood Chit program *[Joint Publication 3-50, Personnel Recovery]*, the United States Government promises to compensate anyone who assists an American service member or other Department of Defense personnel to survive, evade, resist, or escape in hostile territory and return to friendly control. . . . U.S. Government policy is to compensate those individuals who risk their lives, livelihoods, or freedom to assist American service members.

The Legacy of Air America

All American Warriors have heard about ***Air America***. Maybe they have seen the *Air America* motion picture which debuted in 1990 starring Mel Gibson and Robert Downey Jr. Vaguely based upon a detailed and factual book, the action-adventure movie was aimed at a teenage audience. Although entertaining for adolescent boys, it was actually a harmless satirical romantic romp without any resemblance to reality.

> American Warriors have heard about ***Air America***.

Perhaps an interested American Warrior has read the critically acclaimed book that started much of the hoopla. It was first published in the United Kingdom in 1979 under the original title, *The Invisible Air Force: The True Story of the CIA's Secret Airlines*. In 1985, Avon Books republished this riveting saga in the United States under a new title, *Air America*.

Other warriors have heard whispers, rumors, and deadwood talk from old-timers. They've heard tales of getting "sheep-dipped," tales of exchanging the military uniform for civilian clothes in order to fight a secret war for freedom overseas. Where does the truth lie?

The Facts: Born in China in 1947 and growing through the 1940s, 1950s, 1960s, and early 1970s, Air America became the world's largest airline. No other airline owned more planes, and no other airline had bigger or better maintenance facilities. Operating under several names, Air America provided (1) passenger service, (2) maintenance, and (3) military "black ops" for the United States. The airline enjoyed unlimited financial backing because the U.S. Central Intelligence Agency (CIA) funded and directed its entire operation.

In May 1966 an Air America C-47 transport (note "Air America" painted on the fuselage) loaded with rice and weapons waits by the dirt airstrip at Dong Ha in South Vietnam. Although the Air America aircrews used U.S. military facilities, they were paid and directed solely by the CIA (photo by the author, Marion Sturkey).

For the layman, background information is in order. Years before the United States entered World War II in December 1941, Japan had invaded hapless China. Although desperate to thwart the Japanese, the United States was not willing to *officially* enter the war. Therefore, the American government financed and equipped Claire L. Chennault and his "American Volunteer Group" and sent them to China. These *civilian volunteers*, the famous "Flying Tigers," flew their P-40 Warhawk fighters against the Imperial Japanese Air Force during a bloody struggle for the Chinese mainland.

Years before the United States entered World War II in December 1941, Japan had invaded hapless China.

World War II *officially* ended when the Japanese surrendered in 1945.

For the United States, World War II *officially* ended when the Japanese surrendered in 1945. However, the battle continued in China as Chiang Kai-shek and his nationalist armed forces battled a growing communist army led by Mao Tse-tung. To *unofficially* assist its Chinese allies, the United States used the American Volunteer Group organization. By that time it was known by its airline name, Civil Air Transport, commonly called CAT. The Chinese government was its only customer. In later years a noted historian would explain:

> By the end of 1948, American CAT mercenaries were deeply involved in China's bloody civil war.

The new airline operated a fleet of surplus U.S. military transport planes to support the war machine of the Chinese nationalists. CAT moved its headquarters to the island fortress of Taiwan in 1950 after Free China set up its government there. That year the U.S. Central Intelligence Agency (CIA) *acquired* CAT as the first of a worldwide network of "air priority" assets. One U.S. technical report explained the secret acquisition:

> The real story of Air America began in 1950 when the CIA decided that it required an air transport capability to conduct covert operations in Asia.

Operating under a variety of names, including Air America, Air Asia Ltd., and Southern Air Transport, CAT established a gigantic maintenance base at Tainan City, Taiwan. At its peak in the late 1960s the facility would employ over 8,000 people, maintain its own planes, and perform contract maintenance for the U.S. Armed Forces. As cover for covert flights in China and elsewhere, CAT inaugurated civilian passenger service to Tokyo, Bangkok, Manila, and other locations in the Far East.

> CAT established a gigantic maintenance base at Tainan City, Taiwan.

During the Korean War in the early 1950s the men of CAT were called upon for a host of clandestine missions.

> The CAT crews supplied CIA projects in Tibet and Burma and flew over 100 secret night missions over mainland China.

In addition to flights in Korea, the CAT crews supplied CIA projects in Tibet and Burma and flew over 100 night missions over mainland China. Most went well, but some did not. For example, on 29 November 1952 an Air America C-47 took off from Seoul to rescue a CIA agent in Manchuria. Chinese forces downed the plane near Antu in Jilin Province. The shoot-down killed both pilots, and the Chinese Army captured the two CIA officers aboard, Richard Fecteau and John Downey (the Chinese jailed the men for 20 years and released them only after Richard Nixon, the U.S. President, visited China in 1972).

When the French Army was on the ropes in Indochina and the U.S. Armed Forces could not intervene, the CIA called upon Air America. By 1953 the mercenary American airmen were flying troops, ammunition, and food for the French military. Soon Air America crews were swooping into the firestorm at Dien Bien Phu. By the time that struggle reached its peak in early 1954, only one-fourth of the military flights in Indochina were flown by French military aircrews. Air America flew all the others. For public consumption these American mercenaries never existed, but the secrets began to trickle out. In his internationally acclaimed *Hell in a Very Small Place* the French historian, Dr. Bernard B. Fall, explained:

> The U.S. Government consolidated its air priority assets under a single name, ***Air America***.

> Twenty-four of the twenty-nine C-119s flying as part of the French supply operations had American crews under contract to the Taiwan-based Civil Air Transport.

In January 1967 two Marine Corps CH-46 helicopters (first helicopter depicted), with all external identification painted-over for secrecy, fly Chinese Nung mercenaries toward a clandestine drop-off point west of Ban Houe Sane in Laos. The Nungs were hired killers, paid "by the ear" by the CIA. These secret cross-border flights into Laos supported Air America and CIA projects, but the Marine Corps and Army aircrews were paid and directed by the U.S. Department of Defense, not the CIA (photo by the author, Marion Sturkey).

Communist forces drove the French Army out of Indochina, but the war in Cambodia, Laos, and Vietnam heated up again five years later. Western World security interests were deemed at risk, and the U.S. government consolidated its air priority assets under a single name, ***Air America***. The CIA recruited some of its pilots and cargo-kickers from civilian sources, but most mercenary airmen came from the U.S. Armed Forces. Air America made it patently clear whom they wanted: "the most highly skilled, adventurous, and patriotic aviation personnel who could be found."

Those accepted got "sheep-dipped" and vanished from the U.S. Armed Forces.

Skilled military pilots and aircrewmen got an opportunity to apply for reassignment. Those accepted got "sheep-dipped" and vanished from the U.S. Armed Forces. As *civilians* they were spirited away to Southeast Asia where they flew both fixed-wing transports and helicopters. In small black letters on the silver fuselage of each aircraft were two words: Air America. The motto of the clandestine force was simple: "Anything, Anytime, Anywhere." *Hard Rice* was ammunition. *SAR* was an easy way to get yourself killed in Cambodia, Laos, China, North Vietnam, South Vietnam, Thailand, or wherever. It didn't matter. When you don't officially exist there are no restrictions.

The United States acknowledged its involvement in the war in Vietnam, but denied U.S. Armed Forces were fighting elsewhere in Indochina. Yet, the CIA needed military support for its Air America projects in embattled Cambodia and Laos. On *black* operations with names like Project Delta and Shining Brass, U.S. Army and U.S. Marine Corps pilots began flying mercenaries into Laos. They raided North Vietnamese Army (NVA) bases, ambushed NVA units, and then melted back into the Laotian jungle. Capt. Joseph D. "Joe" Snyder, a Marine pilot in squadron HMM-161, wrote in March 1969:

> When you don't officially exist there are no restrictions.

> Another good thing we're doing is inserting mercenaries into Laos. There are some weird things going on over here. The poor folks back home have no idea!

> Officially, no American ever flew into Laos. Whenever anyone turned up dead or missing in that forbidden neverland, the casualty was simply coded to "Project Delta" with no further explanation. . . . On these secret flights into Laos we all knew we were playing a deadly game without any rules. If we got shot down, our wingman would try to swoop in and pick us up. If that failed, we would try to "escape and evade" on foot and make our way back to an American outpost in South Vietnam.
>
> Marion Sturkey, writing in *BONNIE-SUE: A Marine Corps Helicopter Squadron in Vietnam*

For the United States the war in Indochina ended in 1975, and Air America more or less disbanded. Most former military sheep-dipped pilots and aircrewmen had survived. They eventually surfaced, somehow slipping back through the looking-glass into the Air Force, Navy, Marines, Army, or wherever they had vanished from, always with no questions asked. Most of these patriots joined a new organization, the Air America Club. Official U.S. government recognition of Air America, which had never before existed, would finally arrive a decade later in Texas.

The ***Air America Memorial*** was unveiled at McDermott Library on the University of Texas campus at Dallas. William E. Colby, former CIA Director, dedicated the memorial on 30 May 1987. He titled his dedication address, "Courage in Civilian Clothes." Starting alphabetically with the name Robert P. Abrams, the bronze memorial lists names of 242 civilian patriots who lost their lives while flying for Air America. The inscription begins:

> This memorial is dedicated to the aircrews and ground support personnel of Civil Air Transport, Air America, Air Asia, and Southern Air Transport, who died while serving the cause of freedom in Asia from 1947 to 1975.

William P. Clements Jr., Governor of Texas, offered his greetings to Air America survivors, family members, and friends He told them:

> It is important to the health and welfare of our nation to remember these veterans and volunteers who answered the call of duty. A nation cannot ask more of any citizen than to [ask him to] risk his life for the country's preservation.

Ronald Reagan, the U.S. President, extended his appreciation and best wishes via a personally-signed letter to each survivor:

The White House
Washington

. . . The unique service you shared in defense of freedom forged a bond of brotherhood that time and distance cannot break. Unsung and unrecognized, each of you confronted danger and endured terrible hardships, and each of you rose to the challenge; you never faltered. Although free people everywhere owe you more than we can hope to repay, our greatest debt is to your companions who gave their last full measure of devotion. Just as their names are inscribed on this memorial, so their memories are inscribed on our hearts. We will never forget them or their families. God bless you, and God bless America.

[signed] Ronald Reagan

Did Air America, by any name, really fade away? The American public eventually learned that Air America had flown in Guatemala, Angola, the Bay of Pigs – and no one was admitting where else. The U.S. intelligence community remained on the job. Somewhere in the world there always seemed to be a need for clandestine air priority assets. In the introduction to *The Invisible Air Force: The True Story of the CIA's Secret Airlines*, author Christopher Robbins wrote:

Air America is a company incorporated in Delaware, but it is also a generic name used to describe all the CIA air activities . . . . There is a web of dozens of CIA airlines throughout the world which should go under the title, ***CIA Air***. But, that is a logo you will not find anywhere.

From time to time the wall of secrecy breaks down.

From time to time the wall of secrecy breaks down. In the 1980s in Nicaragua the Contras, CIA-backed anticommunist rebels, battled the Sandinista government. On 5 October 1986 the Sandinistas shot down a big gun-running C-123 transport plane. In the wreckage were over 50,000 rounds of ammunition, hundreds of AK-47 rifles, and crates of RPGs. Documents found in the rubble revealed the plane was operated by Southern Air Transport – sound familiar?

The crash had killed the pilots, William Cooper and Wallace Sawyer. The only survivor was Eugene Hausenfauss, USMC veteran, and he told his captors he worked for the American CIA. Active duty LtCol. Oliver North, USMC, had been assigned to the U.S. National Security Council. In televised testimony before the U.S. Congress this patriot would later acknowledge his role as coordinator for covert deliveries of money and arms to the Contras.

Some may wonder whether or not Air America still exists and flies today.

Some may believe the shoot-down in Nicaragua was a fluke. Yet, 24 hours earlier a *civil* four-engine Lockheed L-382 transport had crashed during a nighttime takeoff at Kelly AFB in Texas. The crash happened on a (1) U.S. military base, and the cargo was (2) military equipment on a (3) "military cargo" flight, but the *civil* NTSB investigated the accident (see NTSB /AAR-87/04 for details). The three deceased crewmen were *civilians*, and the L-382 transport (a stretched variant of the big military C-130 Hercules) was owned by Southern Air Transport – once again, sound familiar? In 2008 a technical report explained bluntly:

The accident happened on a (1) military base, and the cargo was (2) military equipment, on a (3) "military cargo" flight.

> Southern Air Transport (SAT), based in Miami, Florida, is best known as a front company for the Central Intelligence Agency. It was founded in 1947 and became a subsidiary of the CIA's airline proprietary network.

SAT filed for bankruptcy on 11 October 1998, the day when the CIA released a report critical of its operation. Only five months later a new corporate entity – a new airline – acquired the SAT planes and other assets. The new airline's web site soon openly stated that it specialized in "cargo transportation for the U.S. Military."

Some may wonder whether or not Air America still exists and flies today. By the Air America company name, the answer is "no." The *original* Air America now flies only in the hearts of men who manned the planes several decades ago. Their contributions to the cause of freedom remain unparalleled in the annals of aviation history. They are *civilian* American Warriors of the long Cold War.

The *original* Air America now flies only in the hearts of men who manned the planes several decades ago.

However, today in the Twenty-First Century the U.S. intelligence community still has a need for air priority assets. The struggle against international terrorism has created a demand for covert aerial *Black Ops* and *civilian patriots* to man the planes. Any person with computer savvy can identify international contract air carriers. With a little time and a little more savvy, any person can identify a carrier's government customers. And, if a person knows the right sources it's quite easy to determine the type of cargo, its origin, and its destination.

Today there are *new* flight crews, *new* planes, and *new* corporate names.

Today there are *new* flight crews, *new* planes, and *new* corporate names. The need remains, and the mission continues. The legacy of Air America will never fade away.

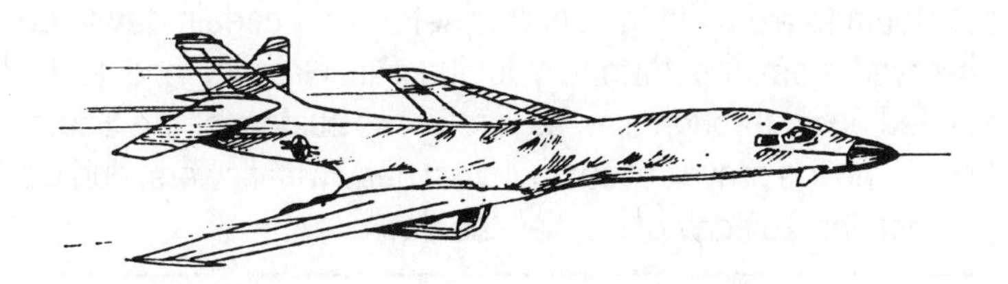

Days of Honor and Marble Tombs

– Memorial Day –

During and after the bloody American Civil War (1861-1865), citizens often placed flowers on graves of soldiers who had fallen in battle. One of these *decorations* of graves took place in Mississippi on 25 April 1866. Women went to a local cemetery and placed flowers on graves of Confederate soldiers killed in 1862 at the Battle of Shiloh in Tennessee. In the same cemetery they saw bare and neglected graves of unidentified Union soldiers, and they placed flowers on those graves as well.

> Citizens often placed flowers on graves of soldiers who had fallen in battle.

In Columbus, Georgia, a military widow had been placing flowers on her husband's grave each day for several years. Encouraged by such loyalty, citizens in Columbus helped her decorate graves of other soldiers who had been killed. In March 1868 the widow wrote to the local *Columbus Times* newspaper with a request:

> We beg the assistance of the press and the ladies throughout the South to aid us in the effort to set apart a certain day to be observed from the Potomac to the Rio Grande, and to be handed down through time as a religious custom of the South, to wreathe the graves of our martyred dead with flowers, and we propose the 26th day of April as the day.

The concept of a special day to place flowers on graves of soldiers killed-in-action appealed to many. In Washington, DC, Gen. John A. Logan commanded the Grand Army of the Republic, a U.S. Armed

Forces veterans organization. On 5 May 1868 he issued General Order No. 11, which began:

> The 30th day of May, 1868, is designated for the purpose of strewing with flowers or otherwise decorating the graves of comrades who died in defense of their country during the late rebellion, and whose bodies now lie in almost every city, village, and hamlet churchyard in the land. In this observance no form of ceremony is prescribed, but posts and comrades will in their own way arrange such fitting services and testimonials of respect as circumstances will permit.

On 30 May of that year in Arlington National Cemetery, citizens recited prayers, sang hymns, and placed flowers on graves of Union and Confederate soldiers. This practice spread across America and became known as Decoration Day. Gradually that name fell out of favor and was replaced by ***Memorial Day***, and the U.S. Congress finally made it a federal holiday in 1967. Congress expanded this day of honor to include (1) all branches of the U.S. Armed Forces and (2) all wars and armed conflicts in which American Warriors have sacrificed their lives for their country.

> The concept of a special day to place flowers on graves of soldiers killed-in-action appealed to many.

The next year, 1968 (the height of the hippie and flower-power movement), Congress passed the Uniform Monday Holiday Act. This new law arbitrarily moved four federal holidays (Washington's Birthday, Veterans Day, Memorial Day, and Columbus Day) to specified Mondays. This was done to create three-day weekends for federal employees. Only federal government employees, including all federal elected officials, were affected. States are not required to observe federal government holidays unless they *elect* to do so. Many

> Citizens recited prayers, sang hymns, and placed flowers at graves of Union and Confederate soldiers.

On 15 September 1950 at Inchon, Korea, infantrymen from Third Platoon, Company A, First Battalion, Fifth Marines, use wooden scaling ladders to clamber over the seawall for their costly assault on North Korean Army bunkers (photo courtesy of U.S. Marine Corps).

states ignored the federal Memorial Day change. One angry citizen lambasted the selfish self-serving congressional decision, and he lamented to the print and broadcast news media:

> Changing the date merely to create a three-day weekend has undermined the meaning of the day.

In spite of congressional malfeasance, Americans continued to honor their war dead. According to tradition, Memorial Day is observed by placing flowers or small flags on graves of patriots who fell in battle. Citizens are encouraged to visit military memorials and fly the American flag at half-mast until noon. They also are asked to fly the POW/MIA flag approved under the 1998 Defense Authorization Act. Citizens observe a "Moment of Remembrance" at 1500 Hours and pledge to aid families of the honored dead.

On the Thursday before Memorial Day, soldiers of the Third U.S. Infantry Regiment place small American flags at each of the quarter-million-plus graves in Arlington National Cemetery. The soldiers then patrol the cemetery through Memorial Day to make sure nobody disturbs the flags. In like manner at St. Louis, Missouri, Boy Scouts and Cub Scouts place flags at over 100,000 graves in the Jefferson Barracks National Cemetery. On Memorial Day, beginning in 1998, Boy Scouts and Girl Scouts began placing candles at graves of soldiers buried in the national military park on Marye's Heights in Virginia. And in 2004, for the first time in over 60 years the city of Washington, DC, sponsored a Memorial Day parade.

Boy Scouts and Cub Scouts place flags at over 100,000 graves in the Jefferson Barracks National Cemetery.

For the first time in over 60 years the city of Washington, DC, sponsored a Memorial Day parade.

– Confederate Memorial Day –

By law, on whatever date, Memorial Day is a federal and state holiday in all 50 states. In southern states in addition to Memorial Day, citizens observe a *state* holiday, ***Confederate Memorial Day***. On this day they honor warriors of the Confederate States of America who died in battle during the American Civil War, commonly known in southern states as the War of Northern Aggression or the War for Southern Independence. Confederate Memorial Day (in Texas, called *Texas Heroes Day*) is a *state* holiday, so each state has selected its own day of observance. For example, the South Carolina Code, Title 53, Chapter 5, Section 53-5-10, states in part:

> Legal holidays enumerated; state employees: . . . the tenth day of May, Confederate Memorial Day.

Confederate Memorial Day is observed in Florida, Georgia, and Mississippi on 26 April; in South Carolina and North Carolina on 10 May; and in Alabama and Mississippi on the fourth Monday in April. It is observed in Virginia on the last Monday in May (same day as the federal holiday); in Kentucky, Tennessee, and Louisiana on 3 June (birthday of Jefferson Davis, CSA President); and in Texas and Arkansas on 19 January (birthday of Gen. Robert E. Lee, CSA).

– Veterans Day –

After four long years of horror and carnage in Europe the giant cannons mercifully fell silent. At 1100 Hours on 11 November 1918 an armistice signed by the Allies and Germany took effect. The Great War, often called The World War and also The War to End All Wars, came to an end. The world rejoiced, and thereafter 11 November became ***Armistice Day*** in most countries (in Canada it is *Remembrance Day*). Woodrow Wilson, the U.S. President, commemorated Armistice Day with these words:

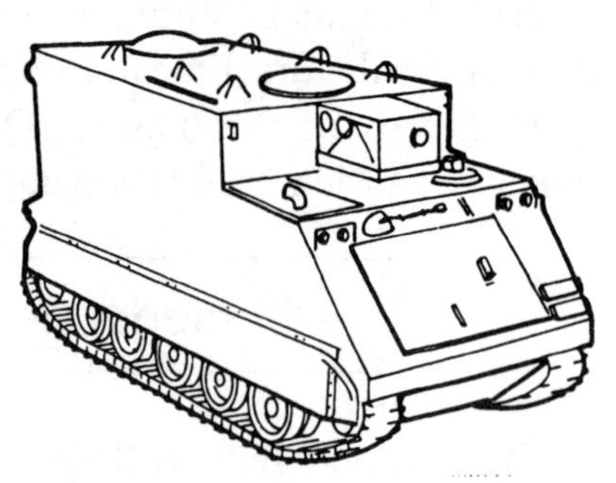

> To us in America, the reflections of Armistice Day will be filled with solemn pride in the heroism of those who died in the country's service and with gratitude for the victory, both because of the thing from which it has freed us and because of the opportunity it has given America to show her sympathy with peace and justice in the councils of nations.

The unofficial protocol for observing Armistice Day called for parades, picnics, barbeque, prayer, and suspension of all business on 11 November each year. In 1926 the U.S. Congress recognized the end of The Great War with a resolution:

> Be it resolved by the Senate – the House of Representatives concurring – that the President of the United States is requested to issue a proclamation calling upon officials to display the flag of the United States on all government buildings on November 11 and inviting the people of the United States to observe the day in schools, churches, or other suitable places . . . .

In 1938 the U.S. Congress made Armistice Day a federal holiday, a day to honor veterans of The Great War, a day "dedicated to the cause of world peace." However, the 1940s witnessed the greatest military mobilization in history for what was called World War II. Five years later American Warriors were again called upon, this time to battle aggression in Korea. Consequently, in 1954, Congress changed the name of Armistice Day to ***Veterans Day***. The U.S. President, Dwight D. Eisenhower, issued a proclamation encouraging "all veterans, all veterans' organizations, and the entire citizenry" to join together in observing this day of honor. The proclamation stated in part:

In 1954, Congress changed the name of Armistice Day to ***Veterans Day***.

> Let us solemnly remember the sacrifices of all those who fought so valiantly, on the seas, in the air, and on foreign shores, to preserve our heritage of freedom, and let us reconsecrate ourselves to the task of promoting an enduring peace so that their efforts shall not have been in vain.

Congress passed the Uniform Monday Holiday Act in 1968. This legislation, effective in 1971, moved Veterans Day and three other federal holidays to a Monday in order to give federal employees four three-day weekends. In 1971 the first Veterans Day observance under the new law created mass confusion. The

In 1971 the first Veterans Day observance under the new law created mass confusion.

federal holiday fell on Monday, 25 October. Yet, most Americans observed 11 November, the date on which the armistice had been signed to end The Great War. Most citizens refused to change.

In addition, most states refused to change their *state* Veterans Day holiday date. Finally bowing to public pressure (something it rarely does) in 1978, Congress changed the Veterans Day federal holiday back to 11 November. Thereafter the meaning and significance of Veterans Day no longer could be lost in a routine three-day weekend. So, each year on the anniversary of the armistice in 1918, America honors all her military veterans, living and dead, who served in the Armed Forces in time of war.

Congress changed the Veterans Day federal holiday date back to 11 November.

– Tomb of the Unknown Soldier –

Shortly after The Great War (today called *World War I*) ended a British Army chaplain, Rev. David Railton visited a military cemetery in France. He knelt and prayed beside graves of British soldiers killed in battle. He saw a grave marked by a wooden cross upon which was handwritten: "An unknown British soldier of the Black Watch." Rev. Railton later wrote about the unidentified soldier:

He knelt and prayed beside graves of British soldiers killed in battle.

> Who was he, and who were [his relatives and friends]? Was he just a laddie? There was no answer to these questions, nor has there ever been, so I thought, [and I received] this answer, clear and strong. Let this body – this symbol of him – be carried reverently over the sea to his native land.

The British government agreed, and amid much ceremony the unknown soldier was disinterred in France and brought home to

England. In the presence of King George V, he was reinterred in Westminster Abbey on Armistice Day in 1920. The initial tomb inscription, later to be supplemented, read in part:

A British Warrior Who Fell In The Great War,
1914-1918, For King And Country.
Greater Love Hath No Man Than This.

Congress approved reinterrment of an "American unknown soldier."

Another such gesture of honor took place in France. An unidentified French combatant killed in The Great War was reinterred at the Arc de Triomphe. The commanding general of American forces in France had learned of the English and French projects while they were in the planning stages. He proposed a similar plan to the U.S. Army Chief of Staff, and in March 1921 the U.S. Congress approved reinterrment of an "American unknown soldier."

The remains of an unidentified American combatant – there were hundreds from which to choose – were disinterred in France. After returning to the United States aboard the *USS Olympia* he lay in state in the U.S. Capitol Rotunda until Armistice Day in 1921. On that day his remains were reinterred in a marble tomb on the plaza of the Memorial Amphitheater in Arlington National Cemetery. Warren G. Harding, U.S. President, officiated at the state funeral. The tomb has never officially been named, but it is most often called the ***Tomb of the Unknown Soldier***. The marble sarcophagus is inscribed:

He lay in state in the U.S. Capitol Rotunda until Armistice Day

HERE RESTS IN
HONORED GLORY
AN AMERICAN
SOLDIER
KNOWN BUT TO GOD

In falling snow, a sentinel from the Third U.S. Army Regiment walks his post at the Tomb of the Unknown Soldier in Arlington National Cemetery (photo courtesy of U.S. Air Force).

On Memorial Day in 1958 two more unidentified American war dead were interred beside their World War I compatriot. One had been killed in World War II, and the other had lost his life in the service of his country during the war in Korea. Twenty-six years later in 1984 an unidentified American Warrior killed during the Vietnam War was interred at the tomb. He would remain there only 14 years, for his remains were exhumed in 1998. With use of DNA analysis and other modern identification protocols he was identified as Lt. Michael J. Blassie, USAF. His family accepted his remains and buried them in a family cemetery plot.

> Since 1937 the tomb has been guarded 24 hours per day, 365 days per year, by sentinels from "The Old Guard," the Third U.S. Army Regiment.

The three unidentified combatants still interred at the tomb now represent all missing and unidentified American Warriors, from all wars, who have fallen in battle in the service of their country. Many families whose son, father, uncle, husband, or brother never returned home from conflict abroad make frequent visits to the Tomb of the Unknown Soldier. Since 1937 the tomb has been guarded 24 hours per day, 365 days per year, by sentinels from "The Old Guard," the Third U.S. Army Regiment. Inclusion in this elite unit is one of the highest honors the Army can bestow upon a soldier. The ritual "changing of the guard" each hour during the day, and each two hours at night, is among the most solemn military ceremonies in the world.

> A bugler sounds *Taps*. A grateful nation has not forgotten.

Each year on Veterans Day and Memorial Day the focal point for national observance is the Tomb of the Unknown Soldier. At 1100 Hours a color guard, which includes members from all branches of the U.S. Armed Forces, executes *Present Arms*. After prayers and a eulogy the President of the United States lays a wreath upon the tomb, steps back, and salutes. A bugler sounds *Taps*. A grateful nation has not forgotten.

– Tomb of the Unknown – Revolutionary War Soldier

William Penn had laid out a blueprint for the city of Philadelphia, Pennsylvania, in 1682. One of the huge public parks designated by Penn was used as a graveyard for American soldiers during the Revolutionary War (1775-1783). Thousands of soldiers who died from wounds and disease were buried within the park in mass graves. After the war ended the city and its residents used the park as a cow pasture, potters field, fishing hole, and often as a religious revival meeting ground. In 1825, Philadelphia officially named the park Washington Square in honor of Gen. George Washington, the Colonial Army commander-in-chief and the first U.S. President.

> Thousands of soldiers who died from wounds and disease were buried within the park in mass graves.

Over 100 years later the city built a stone and bronze memorial to honor (1) Gen. Washington and (2) an "unknown soldier" of the Revolutionary War. In 1956 an archeological team began searching for the unmarked mass graves filled with soldiers. After unearthing nine paupers the team discovered a military mass grave. Remains were verified and selected based upon metal remnants of the soldier's uniform and skeletal damage caused by a musket ball.

The ***Tomb of the Unknown Revolutionary War Soldier*** now stands in Washington Square. A bronze statue depicts Gen. Washington gazing toward nearby Independence Hall. Under an eternal flame the unidentified soldier lies in a tomb inscribed in part:

> Beneath this stone rests a soldier of Washington's Army who died to give you liberty.
>
> In unmarked graves within this square lie thousands of unknown soldiers of Washington's Army who died of wounds and sickness during the Revolutionary War.

There is no sentinel or guard at the tomb in Washington Square. However, throughout each year it is the site of ceremonial functions and a host of military observances.

– Tomb of the Unknown – Confederate Soldier

The Confederate States of America was comprised of 11 southern states that seceded from the United States of America. Born in 1861, the Confederacy perished four years later. At the outset of the American Civil War (also called the *War for Southern Independence*, or the *War of Northern Aggression*, or the *War to Save the Union*, depending upon one's viewpoint), the manufacturing capacity of the agricultural South had been only 11 percent that of the industrialized North. The South was outnumbered roughly six-to-one, and support from Europe never materialized. By early 1865 its industry and infrastructure lay in ruins. The famed Confederate Armed Forces, out of munitions and supplies, finally succumbed to a Union Army with vastly greater manpower and resources. By then Union forces had sacked and burned most major southern cities, including the capital at Richmond, Virginia. The Confederate president and his cabinet members had been captured and imprisoned – the war was over.

> Born in 1861, the Confederacy perished four years later.

Over 100 years later the remains of an unidentified Confederate soldier, authenticated by artifacts accompanying his remains, were discovered on a battlefield site near Vicksburg, Mississippi, in 1979. By that time, of course, the Confederacy existed only in historical lore and in the hearts of ancestors of her former citizens. Yet, several organizations including (1) Sons of Confederate Veterans, (2) United Daughters of the Confederacy, and (3) Military Order of the Stars and Bars began a campaign to honor the fallen unknown soldier and all other unidentified Confederate war dead.

The ***Tomb of the Unknown Confederate Soldier*** was dedicated 6 June 1981 in Biloxi, Mississippi, on Beauvoir (meaning, *beautiful view*), the Jefferson Davis Home and Presidential Library (Jefferson Davis was the CSA President). The tomb is inscribed with the Great Seal of the Confederate States of America; a poem by Father Abram J. Ryan, poet-priest of the Confederacy; and these words:

> The Unknown Soldier
> of the
> Confederate States Of America
> Known But To God

Postscript: During the war the CSA Congress had authorized medals to honor valor in battle. Criteria for the CSA Medal of Honor were almost identical to criteria for the U.S. Medal of Honor: intrepidity, gallantry, and risk of one's life above and beyond the call of duty. Although the CSA Medal of Honor had been approved, none had been awarded during the war.

The CSA Congress had authorized medals to honor valor in battle.

During a formal ceremony in May 1983 the Sons of Confederate Veterans organization posthumously awarded the CSA Medal of Honor to the Unknown Confederate Soldier. Today this one-of-a-kind medal is reverently displayed inside the museum at Beauvoir.

Jefferson Davis (1808-1889), President, Confederate States of America (image courtesy of U.S. National Archives and Records Administration).

Combat Axioms for American Warriors

Since time immemorial the professional warrior has remained a key player on the world stage. Non-combatants look at warriors and perceive either glamor and glory, or horror and hardship. Yet, down through the centuries the warrior has not concerned himself with what non-combatants believe. Instead, the warrior has three concerns: (1) loyalty to brothers-in-arms, (2) allegiance to cause, and (3) success in battle – often called *staying alive* in battle.

Success in battle, like success in other human endeavors, hinges on proven principles. These basic principles do not change. Weapons change, and military technology changes. Nations rise, and nations fall. Causes come, and causes go. Yet, the basic principles of warfare remain constant.

Success in battle, like success in other human endeavors, hinges on proven principles.

Below are selected statements and words of advice to benefit the professional American Warrior – ***fighting words for fighting men***. Most come from warriors who have tasted the sting of battle. Others come from poets, philosophers and heads of state. Readers will find no tomfoolery from pacifists, no edicts from bean counters. There are no theories from REMFs, no sanctimonious advice from brain-dead liberals, no directives from in-the-rear-with-the-gear staff pogues. Instead, readers will find a unique collection of ***combat axioms***, which are displayed chronologically. The wise warrior will read them, heed them, and survive in battle:

The basic principles of warfare remain constant.

Our business in the field of fight
Is not to question, but to fight.
[Homer, *The Iliad*, c 800 BC]

All warfare is based on deception. *[and also]* Invincibility lies in the defense; the possibility of victory in the attack. One defends when his strength is inadequate; he attacks when it is abundant.
[Sun Tzu, *The Art of War*, c 500 BC]

In war, opportunity waits for no man.
[Pericles (c 495-429 BC), Athenian general and statesman]

Great deeds are usually wrought at great risk.
[Herodotus (c 484-425 BC), Greek historian]

Great deeds are usually wrought at great risk.

There is nothing like the sight of an enemy down on his luck.
[Euripides (c 480-406 BC), Greek poet and playwright]

The bravest are surely those who have the clearest vision of what is before them, glory and danger alike, and yet notwithstanding, go out to meet it.
[Thucydides (c 460-395 BC), Greek historian]

Stand firm, for well you know that hardship and danger are the price of glory!
[Alexander the Great (356-323 BC), during a battle in India]

The Spartans do not ask how many the enemy number, but where they are.
[Agis IV, King of Sparta, c 242 BC]

The greater the difficulty, the greater the glory.
[Cicero (106-43 BC), Roman Consul]

I came. I saw. I conquered.
[Julius Caesar, Roman General, entire text of his dispatch to the Roman Senate after his victory at Zela, 47 BC]

The body of a dead enemy always smells sweet.

Let them hate us as long as they fear us.
[Caligula (12-41 AD), Roman Emperor]

The body of a dead enemy always smells sweet.
[Aulus Vitellius, Roman Emperor, at Beariacum, 69 AD]

Cowards die many times before their deaths.

Let him who desires peace prepare for war.
[Vegetius, Roman military theorist, c 400 AD]

The greatest happiness is to vanquish your enemies.
[Genghis Khan (c 1162-1227), Mongol conqueror]

The infantry must ever be regarded as the very foundation and nerve of an army.
[Niccolo Machiavelli (1469-1527), *Discourses*]

It is fighting at a great disadvantage to fight those who have nothing to lose.
[Francesco Guiciardini, *Storia d'Italia*, 1564]

Let them hate us as long as they fear us.

Cowards die many times before their deaths;
The valiant never taste of death but once.
[William Shakespeare (1564-1616), *Julius Caesar*]

We few, we happy few, we band of brothers;
For he today that sheds his blood with me
Shall be my brother.
[William Shakespeare (1564-1616), *Henry V*]

A man-o-war is the best ambassador.
[Oliver Cromwell (1599-1658), Lord Protector of Ireland]

He who tries to defend everything defends nothing.

The first blow is half the battle.
[Oliver Goldsmith, *She Stoops to Conquer*, 1773]

Battles are won by superiority of fire. *[and also]* He who tries to defend everything defends nothing.

[Frederick the Great (1712-1786), *Military Testament*]

The battle, sir, is not to the strong alone. It is to the vigilant, the active, the brave.

[Patrick Henry, American statesman, addressing the Virginia Convention of Delegates, 23 March 1775]

We must all hang together, or we shall all hang separately.

[Benjamin Franklin, American statesman, after signing the Declaration of Independence, 4 July 1776]

The bayonet has always been the weapon of the brave and the chief tool of victory.

If we desire to avoid insult, we must be able to repel it. If we desire peace, it must be known that we are ready for war.

[Gen. George Washington (1732-1799), Continental Army]

Let us beware of being lulled into a dangerous security of being weakened by internal contentions and diversions; of neglect in military exercises and disciplines in providing stores and arms and munitions of war.

[Benjamin Franklin (1706-1790), American statesman]

March to the sound of the guns.

[Duke of York, British noble, 1793]

No military leader has ever become great without audacity.

[Carl von Clausewitz, *Principles of War*, 1812]

Don't give up the ship! Fight her until she sinks!

[Capt. James Lawrence, USN, as he lay mortally wounded aboard his frigate, *USS Chesapeake*, 1 June 1813]

One man with courage is a majority.

[Thomas Jefferson (1743-1826), U.S. President]

Glory may be fleeting, but obscurity is forever. *[and also]* Moral forces, rather than numbers, decide victory. *[and also]* Good infantry is, without any doubt, the sinew of an army. *[and also]* Never interrupt your enemy when he is making a mistake. *[and also]* The bayonet has always been the weapon of the brave and the chief tool of victory. *[and also]* Four hostile newspapers are more to be feared than a thousand bayonets.

[Napoleon Bonaparte, *Maxims of War*, 1831]

The best strategy is always to be strong. *[and also]* Blood is the price of victory.

[MGen. Carl von Clausewitz, *On War*, 1832]

The best strategy is always to be strong.

The bayonet has always been the weapon of the brave and the chief tool of victory.

Our flag still waves proudly from the walls. . . . I shall never surrender nor retreat. . . . I am determined to sustain myself as long as possible and die like a soldier who never forgets what is due to his own honor and that of his country. Victory or Death!

[LtCol. William B. Travis, Texas Volunteer Militia, in his last dispatch from the Alamo, 24 February 1836]

Gone to Florida to fight the Indians. Will be back when the war is over.

[Col. Archibald Henderson, USMC Commandant, in a handwritten note glued to his office door, 1836]

Gone to Florida to fight the Indians. Will be back when the war is over.

On 30 September 1966 an M-60 machinegun team from Third Battalion, Fourth Marines, battles an entrenched enemy force on Nui Cay Tri Mountain, South Vietnam, amid jungle vegetation ripped apart by air strikes and artillery fire (photo courtesy of U.S. Marine Corps).

A great country can not wage a little war.
[Duke of Wellington, in the House of Lords, 16 January 1838]

We should forgive our enemies – but *only* after they have been hanged first.
[Heinrich Heine (1797-1856), German philosopher]

I was too weak to defend, so I attacked. *[and also]* Do your duty in all things. You can not do more. You should never do less.
[Gen. Robert E. Lee (1807-1870), CSA]

I was too weak to defend, so I attacked.

There is a true glory and a true honor – the glory of duty done, the honor of the integrity of principle.
[Gen. Robert E. Lee, CSA, in *Southern Historical Society Papers*]

Get 'em skeered, and keep the skeer on 'em. *[and also]* I always make it a rule to get there first'est with the most'est. *[and also]* In

any fight it's the first blow that counts the most. *[and also]* War is *fighting*, and fighting means *killing*.
[LtGen. Nathan Bedford Forrest (1821-1877), CSA]

War is *fighting*, and fighting means *killing*.

I make no terms. I accept no compromises.
[Jefferson Davis (1808-1889), CSA President]

There! There! There is Jackson! Standing like a stone wall!
[the rallying shout of an unidentified CSA officer at the battle of Bull Run (also called, *First Manassas*) on 21 July 1861, referring to BGen. Thomas J. "Stonewall" Jackson, CSA]

There! There! There is Jackson! Standing like a stone wall!

Success and glory are in the advance. Disaster and shame lurk in the rear.
[MGen. John Pope, USA, in a General Order, 14 July 1862]

Always mystify, mislead, and surprise the enemy. *[and also]* Duty is ours, consequences are God's. *[and also]* What is life without honor? Degradation is worse than death! *[and also]* An enemy routed, if hotly pursued, becomes panic-stricken and can be destroyed by half their number. *[and also]* To move swiftly, strike vigorously, and secure the fruits of victory is the secret of successful war. *[and also]* The business of the soldier is to fight, to find the enemy and strike him, invade his country, and do him all possible damage in the shortest possible time.
[LtGen. Thomas J. "Stonewall" Jackson (1824-1863), CSA]

We will fight them until Hell freezes over, and then we will fight them on the ice.

We will fight them until Hell freezes over, and then we will fight them on the ice.
[unidentified CSA soldier, Gettysburg, 3 July 1863]

We must substitute *esprit* for numbers.
[MGen. J.E.B. "Jeb" Stuart (1833-1864), CSA]

Get your enemy at a disadvantage and never, on any account, fight him on equal terms.
[George Bernard Shaw, *Arms and the Man*, 1894]

What is life without honor? Degradation is worse than death!

My religious belief teaches me to feel as safe in battle as in bed.
[LtGen. Thomas "Stonewall" Jackson, CSA, quoted posthumously in *Stonewall Jackson*, 1898]

I want no prisoners. I wish you to burn and kill. The more you burn and kill, the better it will please me.

Civilize 'em with a Krag.
[motto of U.S. Marines in China during the Boxer Rebellion in 1900, in regard to their Krag-Jorgensen rifles]

Courage is resistance to fear, mastery of fear, not absence of fear.
[Samuel L. Clemens (1835-1910), aka Mark Twain]

Speak softly, and carry a big stick.
[Col. Theodore "Teddy" Roosevelt, USA, U.S. Vice President, 2 September 1901]

I want no prisoners. I wish you to burn and kill. The more you burn and kill, the better it will please me.
[BGen. Jacob H. Smith, USA, in his order to Maj. L.W.T. Waller, USA, in Samar, October 1901]

A man who is good enough to shed his blood for his Country is good enough to be given a square deal afterwards.
[Col. Theodore "Teddy" Roosevelt, USA, U.S. President, 1903]

Those who cannot remember the past are condemned to repeat it.
[George Santayana, philosopher, *The Life of Reason*, 1906]

The essence of war is violence. Moderation in war is imbecility.
[Adm. Sir John Fisher, RN, in a letter, 25 April 1912]

My religious belief teaches me to feel as safe in battle as in bed.

Find the enemy and shoot him down. Anything else is nonsense. *[and also]* The aggressive spirit, the offensive, is the chief thing everywhere in war, and the air is no exception.
[Baron Capt. Manfred von Richthofen "The Red Baron," German Flying Service (80 air-to-air kills in World War I), 1917]

Retreat, Hell! We just got here!

A pacifist is as surely a traitor to his country and to humanity as is the most brutal wrongdoer.
[Col. Theodore "Teddy" Roosevelt, USA, former U.S. President, 27 July 1917]

Retreat, Hell! We just got here!
[Capt. Lloyd Williams, USMC, to the retreating French Army commander who pleaded with him to flee from attacking German Army troops in Belleau Wood, France, 2 June 1918]

Come on, you sons-of-bitches! Do you want to live forever?
[GySgt. Daniel J. "Dan" Daly, USMC, as he led the attack into Belleau Wood, France, 6 June 1918]

Come on, you sons-of-bitches! Do you want to live forever?

The will to conquer is the first condition of victory.
[Marsh. Ferdinand Foch, *Principles of War*, 1920]

War hath no fury like a noncombatant.
[Charles E. Montague, *Disenchantment*, 1922]

To be vanquished and yet not surrender, that is victory.
[Marsh. Josef Pilsudski (1867-1935), Polish Army]

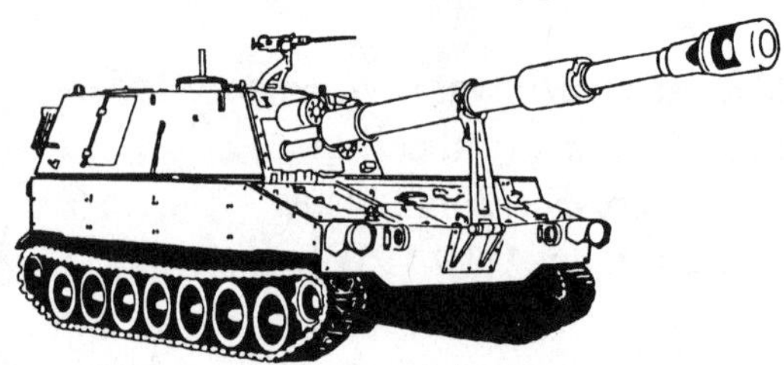

The advantage of sea power used offensively is that, when a fleet sails, one can never be sure where it will strike.
[Sir Winston Churchill, *Their Finest Hour*, 1924]

People sleep peaceably in their beds at night only because rough men stand ready to do violence on their behalf.
[George Orwell (1903-1950), novelist and philosopher]

Through mobility we conquer.
[motto of The Cavalry School, USA, Fort Riley, c 1930]

To wound all ten fingers of a man is not so effective as to chop one of them off. To rout ten of the enemy's divisions is not so effective as to annihilate one of them.
[Mao Tse-tung, Chairman of the People's Republic of China and Commander-in-Chief of the Army, December 1936]

Power emanates from the barrel of a gun.
[Mao Tse-tung, Chairman of the People's Republic of China and Commander-in-Chief of the Army, *On Guerrilla War*, 1938]

Victory at all costs, victory in spite of all terror, victory however long and hard the road may be, for without victory there is no survival.

The more we sweat in peace, the less we bleed in war.
[Vijaya L. Pandit (1900-1990), Indian diplomat]

Victory at all costs, victory in spite of all terror, victory however long and hard the road may be, for without victory there is no survival.
[Sir Winston Churchill, in the House of Commons, 13 May 1940]

We are so outnumbered there's only one thing to do – attack!
[Adm. Andrew Cunningham, RN, at Taranto, 11 November 1940]

> Praise the Lord, and pass the ammunition!

Sure I am of this. You have to endure to conquer. *[and also]* The only thing you must really do is never, never, never, give up. *[and also]* Nothing is worse than war? Dishonor is worse than war. Slavery is worse than war. *[and also]* Battles are won by slaughter and manoeuver.
[Sir Winston Churchill (1874-1965), British Prime Minister]

Praise the Lord, and pass the ammunition!
[Lt. Howell M. Forgy, USN chaplain, to the Navy gun crews aboard *USS New Orleans* during the Japanese air attack at Pearl Harbor, Hawaii, 7 December 1941]

> God favors the bold and the strong of heart.

> Put your heart and soul into being an expert killer.

Put your heart and soul into being an expert killer. The only *good enemy* is a *dead enemy.*
[Gen. George S. Patton Jr., USA, March 1942 (paraphrasing Gen. Philip H. Sheridan, USA, who had stated, "The only good Indian is a dead Indian.")]

God favors the bold and the strong of heart.
[MGen. Alexander Vandergriff, USMC, August 1942]

> War is a bloody, killing business. You've got to spill their blood, or they will spill yours. Rip them up the belly! Shoot them in the guts!

A pint of sweat will save a gallon of blood.
[Gen. George S. Patton Jr., USA, 8 November 1942]

You'll never get a Purple Heart hiding in a foxhole! Follow me!
[Capt. Henry P. Crowe, USMC, Guadalcanal, 13 January 1943]

An MV-22 Osprey in hover mode. The Osprey is a hybrid flying machine that can hover like a helicopter and also fly like a fixed-wing aircraft (photo courtesy of U.S. Marine Corps).

Before we're through with them, the Japanese language will be spoken only in Hell.

[Adm. William F. "Bull" Halsey, USN, 1943]

Casualties many, percentage of dead not known, combat efficiency: we are winning.

[Col. David M. Shoup, USMC, Tarawa, 21 November 1943]

Nobody ever won a war by dying for his country. You win a war by making the ***other*** poor dumb bastard die for ***his*** country. *[and also]* We don't want yellow cowards in this Army. They should be killed off like rats. *[and also]* War is a bloody, killing business. You've got to spill their blood, or they will spill yours. Rip them up the belly! Shoot them in the guts!

[Gen. George S. Patton Jr., USA, addressing soldiers of his U.S. Third Army, in England, 5 June 1944]

Nobody ever won a war by dying for his country. You win a war by making the ***other*** poor dumb bastard die for ***his*** country.

Lead me, follow me, or get out of my way!

[Gen. George S. Patton Jr. (1885-1945) USA]

Darkness is a friend to the skilled infantryman. *[and also]* In war the chief incalculable is the human will.

[Sir B.H. Liddell Hart, British Army, *Thoughts on War*, 1944]

Those poor bastards, they've got us surrounded. Good! Now we can fire in any direction. They won't get away this time!

Wars may be fought with weapons, but they are won by men.

[Gen. George S. Patton Jr., USA, in *The Cavalry Journal*]

The only *good enemy* is a *dead enemy.*

We're not accustomed to occupying defensive positions. It's destructive to morale.

[LtGen. Holland "Howlin' Mad" Smith, USMC, Iwo Jima, 1945]

No sane man is unafraid in battle. But discipline produces in him a form of vicarious courage. *[and also]* To halt under fire is *folly*. To halt under fire, and not fire back, is *suicide*.

[Gen. George S. Patton Jr., USA, quoted posthumously in *War as I Knew It*, 1947]

See, Decide, Attack, Reverse.

[Col. Erich Hartmann, Luftwaffe (352 air-to-air kills, WW II)]

It is fatal to enter any war without the will to win it.

Hit quickly, hit hard, and keep on hitting.

[LtGen. Holland M. Smith, USMC, *Coral and Brass*, 1949]

Hit quickly, hit hard, and keep on hitting.

In war there is never a second-place prize for the runner-up.

[Gen. Omar Bradley, USA, in *Military Review*, February 1950]

Those poor bastards, they've got us surrounded. Good! Now we can fire in any direction. They won't get away this time! *[and also]* Don't forget that you're First Marines! Not all the communists in Hell can overrun you!

[Col. Lewis B. "Chesty" Puller, USMC, when told his regiment was surrounded by seven Chinese Divisions near Chosin Reservoir, Korea, December 1950]

Give me an order to do it. I can break up Russia's five A-bomb nests in a week. And when I go up to meet Christ, I think I could explain to Him that I had saved civilization.

[MGen. Orvil A. Anderson, USAF, 1950]

Diplomacy is the art of saying "nice doggie" until you can find a bigger rock.

It is fatal to enter any war without the will to win it.

[Gen. Douglas MacArthur, USA, 7 July 1952]

War is never prevented by running away from it.

[Air Marsh. Sir John Slessor, RAF, *Strategy for the West*, 1954]

Diplomacy has rarely been able to gain at the conference table what cannot be gained or held on the battlefield.

[Gen. Walter B. Smith, USA, after returning from the Geneva Conference on Korea and Indochina, 1954]

In war there is no substitute for victory.

Success in battle. That is the only objective of military training.

[LtGen. Lewis B. "Chesty" Puller, USMC, 2 August 1956]

It is essential to understand that battles are won primarily in the hearts of men.

[Viscount Field Marsh. Montgomery, British Army, *The Memoirs of Field Marshall Montgomery*, 1958]

A graphic illustration of a new class of aircraft carrier. The nuclear powered behemoth will provide automated damage control functions and an advanced aircraft recovery system (AARS) to reduce crew workload and enhance safety (graphic courtesy of U.S. Navy).

Diplomacy is the art of saying "nice doggie" until you can find a bigger rock.
[Wynn Catlin, Texas political observer]

> It is fatal to enter any war without the will to win it.

There is nothing like seeing the other fellow run to bring back your courage.
[Sir William Slim, British Army, *Unofficial History*, 1959]

Appeasers believe that if you keep on throwing steaks to a tiger, the tiger will become a vegetarian.
[Heywood Brown, American novelist]

I have always regarded the forward edge of the battlefield as the most exclusive club in the world.
[LtGen. Brian Horrocks, British Army, *A Full Life*, 1960]

> Victory is always possible for the person who refuses to stop fighting.

In war there is no substitute for victory. *[and also]* Duty, Honor, Country. These three hallowed words reverently dictate what you *ought* to be, what you *can* be, what you *will* be.
[Gen. Douglas MacArthur, USA, at West Point, 12 May 1962]

Glory is not an end in itself, but rather a reward for valor and faith.
[William J. Bennett, American political observer and educator]

Battles are won primarily in the hearts of men.

Get the blade into the enemy. This is the main principle in bayonet fighting. It is the blade that kills.
[Marine Corps Association, *Guidebook for Marines*, 1962]

The fundamental law of wartime negotiations: you negotiate with the enemy with your knee in his chest and your knife at his throat.
[Gary J. Harris, military theorist]

Victory is always possible for the person who refuses to stop fighting.
[Napoleon Hill, motivational writer]

If you don't fight, you can't win. No guts, no glory.

The only men fit to live are those men who are not afraid to die.
[motto of USN/USMC training squadron VT-5, Pensacola, Florida, 1965]

Find, fix, fight, follow, finish.
[universal military axiom for *destroying* the enemy]

It isn't the size of the dog in the fight that counts. It's the size of the fight in the dog.
[Gen. Dwight D. Eisenhower (1890-1969), USA, U.S. President]

If your bayonet breaks, strike with the stock. If the stock gives way, hit with your fists. If your fists are hurt, bite with your teeth.
[Gen. Makhail Dragomirov, Russian Army, *Notes for Soldiers*]

The only men fit to live are those men who are not afraid to die.

Thirty-caliber-rounds won't hurt you if your mind's right.
[hand-painted wording on a sign in the USMC helicopter pilots' ready-room tent, Phu Bai, Vietnam, 1967]

Answer violence with violence!
[Col. Juan D. Peron (1895-1974), President of Argentina]

Marines defending are like Antichrists at vespers.
[Michael Herr, *Dispatches*, 1968]

If you're in a fair fight, you didn't plan it properly.
[Nick Lappos, Chief R&D pilot, Sikorsky Aircraft Corp.]

A battle plan is good only until enemy contact is made.
[Gen. Norman Schwarzkopf, USA, 1988]

Fight to fly, Fly to fight, Fight to win.
[motto of USN/USMC Fighter Weapons (Top Gun) School]

We should negotiate only when our military superiority is so convincing that we can achieve our objectives at the conference table, and deny the aggressor theirs. *[and also]* The finest steel has gone through the hottest fire.
[Richard M. Nixon (1913-1994), U.S. President]

If you're in a fair fight, you didn't plan it properly.

A Marine's most sought after privilege is to be able to fight for another Marine.
[MGen. Mike Myatt, USMC, Kuwait, 1991]

Nothing in life is more liberating than to fight for a cause larger than yourself.

I love the Corps for those intangible possessions that cannot be issued: pride, honor, integrity, and being able to carry on the traditions for generations of warriors past.
[Cpl. Jeff Sornig, USMC, in *Navy Times*, November 1994]

If you don't fight, you can't win. No guts, no glory.
[anonymous]

We fought for each other, or to uphold the honor of the Corps. That was what mattered.
[Capt. Angus Deming, USMC, in *Newsweek*, 7 August 1995]

God may show you mercy. We will not.

We might succeed, or we might fail, but we would succeed or fail together. There was no other Marine Corps way.
[Capt. Marion F. Sturkey, USMC, *Bonnie-Sue*, 1996]

The purpose of offensive combat is to destroy the enemy and his will to fight.
[Marine Corps Association, *Guidebook for Marines*, 1997]

Nothing in life is more liberating than to fight for a cause larger than yourself.
[Capt. John McCain, USN, former POW in North Vietnam (later U.S. Senator), *Faith of My Fathers*, 1999]

A wise warrior fears the so-called news media, for he knows the sniveling media whores may try to steal his honor.

When America uses force in the world, the cause must be just, the goal must be clear, and the victory must be overwhelming.
[George W. Bush, presidential candidate (later U.S. President), addressing the Republican National Convention on 4 August 2000]

The purpose of offensive combat is to destroy the enemy and his will to fight.

Do not fear the enemy. At the worst, the enemy can only take your life. Instead, a wise warrior fears the so-called news media, for he knows the sniveling media whores may try to steal his honor.
[SSgt. Robert Johnson, USA, Fort Worth, Texas, August 2001]

They should be caught, drawn & quartered, decapitated, and their ugly [expletive] heads put on pikes in front of the White House.
[Maj. Bill F. Weaver, USMC, 12 September 2001]

I say to our enemies, God may show you mercy. We will not.
[Capt. John McCain, USN, also U.S. Senator, in the U.S. Senate, 12 September 2001]

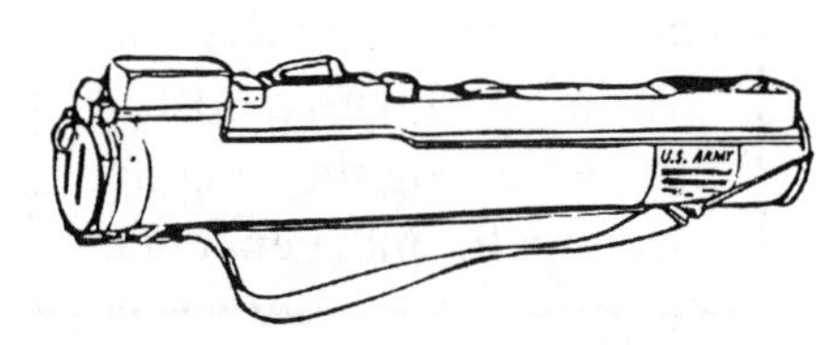

Let the enemy know the full fury of an America that has been wronged and demands retribution. Kill them all!
[Sgt. Arthur W. Larsen, USMC, 13 September 2001]

Shoot at them, and they will kill you. Marines know how to fight. Marines ***like*** to fight.
[Francis J. "Bing" West, USMC, also former U.S. Assistant Secretary of Defense, in *The Wall Street Journal*, 23 July 2003]

Terrorists despise us because we are free, and because we do not believe as they do. We must hunt them down and kill them, all of them. There is no substitute for victory!
[Sgt. Phil P. Lucas, USMC, newsletter editor, in *The Lakelands Leatherneck*, August 2004]

Screw the ACLU, and God bless our warriors!
[U.S. Armed Forces reply to an ACLU complaint alleging that warriors at a memorial service were "government employees" and were "praying on government time," November 2004]

We must hunt them down and kill them, all of them. There is no substitute for victory!

My mom thinks war against terrorism is wrong, and the price is too high. No problem – my buddies and I are paying her share.
[a U.S. Army soldier in the Wounded Warrior ward at the V.A. Hospital, Augusta, Georgia, May 2008]

Patriot Dreams

American Patriot! The words evoke thoughts of Paul Revere, Patrick Henry, and the American Revolution two-hundred-plus years ago. Webster defines *patriot* as "one who loves, supports, and defends his country." Our new fledgling Republic was blessed with a host of patriots during the long-ago struggle for independence. Thomas Jefferson, George Washington, Nathan Hale, John Adams, Thomas Paine, and others gave us government of the people.

Our new fledgling Republic was blessed with a host of patriots during the long-ago struggle for independence.

Those men are now dead and gone. A latter-day philosopher tells us that eternal vigilance is the price of freedom. Yet, American Warriors speak from experience and simply explain: "The price of freedom is not cheap."

American Warriors speak from experience and simply explain: "The price of freedom is not cheap."

History has shown that foreign despots – Lenin, Hitler, Mao, Tito, Stalin, and their henchmen – will always plague the world. Some tyrants like Tojo (1884-1948) earned the hangman's noose. Mussolini (1883-1945) was hung *upside down.* For others the axe would take a long time to fall. Idi Amin (1925-2003), who fed political opponents to crocodiles, survived to live many years in exile with the millions he plundered. Others still rule with an iron hand. Of course, the United States is not immune. Sometimes our own, such as Benedict Arnold (1741-1801), make a mockery of the trust vested in them. Nonetheless, our society has persevered.

He has been called an "extra star on the American Flag."

A one-of-a-kind American icon was born in Winterset, Iowa, in 1907. His parents named him Marion Morrison, but the world would remember him as ***John Wayne*** (1907-1979). He embodied all that is virtuous and good about his native land. He was

a patriot, father, husband, humanitarian, actor, tycoon, role model, but most of all he was a genuine American hero. He has been called an "extra star on the American Flag." He *lived* the virtues that made him a legend. Seventeen years after his death, a Harris Poll revealed that he was then the most popular motion picture actor of all time. John Wayne once explained his love for America:

> Sure, I wave the American flag. Do you know of a better flag to wave? Sure, I love my country with all her faults. I'm not ashamed of that. Never have been. Never will be.

Each generation of Americans has produced patriots willing to serve and protect their country. However, most modern-day patriots aren't recognized as such. They aren't like John Wayne, for most of them are invisible in our society.

Today's patriot is usually the common man, the average citizen, the next-door neighbor, the man who gave his all to his country and asked for nothing in return. He may be the legionnaire who dons his clothes each day with a prosthetic hand. He may be the elderly man bagging groceries at the supermarket, who never speaks of the terror on Iwo Jima in 1945 – he knows nobody still alive who could comprehend such horror. He may be the man who stormed into Afghanistan in 2001. Duty called, and he answered. Other patriots of today are the thousands of warriors who went into battle for their country, and who never returned.

Today's patriot is usually the common man, the average citizen, the next door neighbor, the man who gave his all to his country and asked for nothing in return.

Historically, after diplomacy and reason have failed our country always relies upon its patriots, its military warriors. Although Mao

The American Bald Eagle, adopted as the "official emblem of the United States" in 1782 and often called the "national bird" (photo courtesy of U.S. National War College).

Tse-tung (1893-1976) was no friend of America, he understood international relations and explained: "Power emanates from the barrel of a gun." When enemies and terrorists threaten our country it is always American Warriors, not politicians, who ensure our national survival.

> It is always American Warriors, not politicians, who ensure our national survival.

In March 2001 a twelfth grade schoolgirl in Ohio stumbled across a few vague paragraphs in her history textbook. Only thirty years in the past her country had fought a protracted war in Southeast Asia, she read. Why? What was at stake? Although she was a history buff, the teenager did not know the name of any American involved in that conflict. Intrigued and challenged, she began delving into America's involvement in that war. Her research evolved into her senior thesis, which she titled: "Who Were The Heroes?"

One warrior who fought in Vietnam answered the schoolgirl's public queries. In a long letter he explained that the heroes were just ordinary men. Actually, he said, *boys* would be a better description in most cases. But they were uniquely bound together. They shared a common bond, he explained. They believed in each other, and they believed in their cause. On a personal note the old warrior related two instances wherein his friends had put their lives on the line, above and beyond the call, in an effort to save brothers-in-arms in peril. Then he wrote:

The heroes were just ordinary men.

> The heroes who survived are now in their fifties and sixties. You know them as fathers, uncles, neighbors, maybe teachers. They have jobs and families. They pay taxes and make our society function. They don't label themselves as heroes. Yet, they are American Patriots in every sense of the words. And, deep down inside they still maintain that undying brotherly love for the men with whom they served in Vietnam thirty or forty years ago. Without question, they are your heroes.

Loyalty to brothers-in-arms. Loyalty to cause. Patriotism. These virtues are synonymous with the culture of warriors and patriots. These virtues can't be separated. Throughout history in the civilized world, each generation has produced men who could have taken the easy road through life. Instead they opted for a life of service and sacrifice for society. We should pause and reflect on statements from such men, and about such men. These statements come from warriors, patriots, historians, poets, philosophers, heads of state, and others who understand sacrificial devotion to one's homeland. Their words are arranged in chronological order:

Go tell the Spartans, thou that passeth by,
That here, obedient to the laws, we lie.

[epitaph for Spartan soldiers who fell in battle while holding the pass at Thermopylae in 480 BC]

They opted for a life of service and sacrifice for society.

Who here is so vile that will not love his country?
[William Shakespeare (1564-1616), *Julius Caesar*]

Only a virtuous people are capable of freedom.
[Benjamin Franklin, *Historical Review of Pennsylvania*, 1759]

Is life so dear, or peace so sweet, as to be purchased at the price of chains and slavery? Forbid it, Almighty God! I know not what course others may take, but as for me, give me liberty, or give me death!
[Patrick Henry, American statesman, to the Virginia Convention of Delegates, 23 March 1775]

I only regret that I have but one life to lose for my country.

Stand your ground, men. Don't fire unless fired upon. But if they mean to have a war, let it begin here.
[Capt. James Parker, Colonial Militia, to his "minute men" before the British Army opened fire, 19 April 1775]

Those who expect to reap the blessings of liberty must, like men, undergo the fatigue of supporting it. *[and also]* O ye that love mankind! Ye that dare oppose not only tyranny, but the tyrant, stand forth!
[Thomas Paine, *The American Crisis*, 1776]

... with a firm reliance on the protection of Divine Providence, we mutually pledge to each other our Lives, our Fortunes, and our sacred Honor.
[U.S. Declaration of Independence, 4 July 1776]

If they mean to have a war, let it begin here.

I only regret that I have but one life to lose for my country.
[Capt. Nathan Hale, Continental Army, last words before he was hanged as a spy by the British Army, 22 September 1776]

We fight, get beat, rise, and fight again.
[MGen. Nathanael Greene, Continental Army, 22 June 1781]

The tree of liberty must be refreshed from time to time with the blood of patriots and tyrants.
[Thomas Jefferson, later U.S. President, November 1787]

The only thing necessary for the triumph of evil is for good men to do nothing.

You will never know how much it has cost my generation to preserve your freedom. I hope you will use it wisely.
[John Adams (1735-1826), U.S. President]

The only thing necessary for the triumph of evil is for good men to do nothing.
[Edmund Burke (1729-1797), British statesman]

In matters of principle, stand like a rock!

In matters of principle, stand like a rock! *[and also]* I have sworn, upon the altar of God, eternal hostility against every form of tyranny over the mind of man.
[Thomas Jefferson (1743-1826), U.S. President]

Breathes there a man with soul so dead,
Who never to himself hath said,
This is my own, my native land!
[Sir Walter Scott, *Lay of the Last Minstrel*, 1805]

And the star-spangled banner in triumph shall wave, O'er the land of the free and the home of the brave!
[Francis Scott Key, American attorney and prisoner exchange negotiator, *The Defense of Fort McHenry*, later called *The Star Spangled Banner*, 14 September 1814]

The tree of liberty must be refreshed from time to time with the blood of patriots and tyrants.

Liberty Enlightening the World, a gift to the people of the United States from the people of France, stands in New York harbor. This "Statue of Liberty" was dedicated on 28 October 1886 and became an international symbol of freedom and democracy. It was designated a U.S. National Monument in 1924 (photo courtesy of U.S. National War College).

Our country! In her intercourse with foreign nations, may she always be in the right. But, our country, right or wrong!

[Cmd. Stephen Decatur, USN, at Norfolk, Virginia, 1816]

Let us show ourselves *worthy* to be free, and we *shall* be free.

America is great because she is good. If America ever ceases to be good, America will cease to be great.

[Alexis de Tocqueville, French historian, 1831]

My country, 'tis of thee, sweet land of liberty,
Of thee I sing; land where my fathers died,
Land of the pilgrims' pride, from every mountainside,
Let freedom ring!

[Samuel F. Smith, *America*, 1832]

To the People of Texas and all Americans in the world: . . . I call on you in the name of liberty, patriotism, & of everything dear in the American character, to come to our aid with all dispatch.

[LtCol. William B. Travis, Texas Volunteer Militia, at the Alamo in southwest Texas, 24 February 1836]

All I am, and all I have, is at the service of my Country.

I was born an American,
I live an American,
I shall die an American.

Let us fly to arms, march to the battlefield, meet the foe, and give renewed evidence to the world that the arms of free men, uplifted in defense of liberty and right, are irresistible. Now is the day, now is the hour, when Texas expects every man to do his duty. Let us show ourselves *worthy* to be free, and we *shall* be free.

[Henry Smith, Governor of Texas, 2 March 1836]

One country, one constitution, one destiny. *[and also]* I was born an American, I live an American, I shall die an American. *[and also]* God grants liberty to those who love it and are always ready to guard and defend it.

[Daniel Webster (1782-1852), U.S. Senator and orator]

If anyone attempts to haul down the American flag, shoot him on the spot.

[John A. Dix, U.S. Secretary of the Treasury, 1860]

A thoughtful mind, when it sees a nation's flag, sees not the flag only, but the nation itself.

[Henry Ward Beecher, *The American Flag*, 1861]

All I am, and all I have, is at the service of my Country.
[LtGen. Thomas J. "Stonewall" Jackson, CSA, in a letter, 1861]

One flag, one land, one heart, one hand, one nation, evermore.
[Oliver Wendell Holmes Jr., American jurist, 1862]

Fourscore and seven years ago our fathers brought forth on this continent a new nation, conceived in liberty, and dedicated to the proposition that all men are created equal.
[Abraham Lincoln, U.S. President, at dedication of the National Cemetery at Gettysburg battlefield, 19 November 1863]

Dear Madam: I have been shown in the files of the War Department a statement of the Adjutant General of Massachusetts [which shows] that you are the mother of five sons who have died gloriously on the field of battle. I feel how weak and fruitless must be any words of mine which should attempt to beguile you from the grief of a loss so overwhelming. But I cannot refrain from tendering you the consolation that may be found in the thanks of the Republic they died to save. I pray that our Heavenly Father may assuage the anguish of your bereavement, and leave you only the cherished memory of the loved and lost, and the solemn pride that must be yours to have laid so costly a sacrifice upon the altar of freedom.
[Abraham Lincoln, U.S. President, in a letter to Lydia Bixbey, 21 November 1864 (read verbatim 134 years later in the motion picture, *Saving Private Ryan*, 1998)]

Abandon your animosities and make your sons Americans.

The muster rolls on which the name and oath were written were pledges of honor, redeemable at the gates of death. And those who went up to them, knowing this, are on the list of heroes.
[BGen. Joshua L. Chamberlain, USA, writing in 1866 about USA and CSA volunteers during the American Civil War]

Abandon your animosities and make your sons Americans.
[Gen. Robert E. Lee, CSA, after the battle at Appomattox, 1865]

War is an ugly thing, but not the ugliest thing. The decayed and degraded state of moral and patriotic feelings which thinks that nothing is worse than war is much worse. A man who has nothing for which he is willing to fight, nothing which is more important than his own personal safety, is a miserable creature and has no chance of being free – unless made and kept so by the exertions of better men than himself.
[John S. Mill, British philosopher, 1868]

A man who has nothing for which he is willing to fight, nothing which is more important than his own personal safety, is a miserable creature.

There is something magnificent about having a country to love.
[James Russell Lowell (1819-1891), American poet]

Eternal vigilance is the price of liberty.
[Wendell Phillips (1811-1884), American orator]

A man's country is not a certain area of land . . . it is a principle, and patriotism is loyalty to that principle.
[George William Curtis (1824-1892), American author]

There is something magnificent about having a country to love.

I pledge allegiance to the Flag of the United States of America and to the Republic for which it stands, one nation, under God *[see note, below]*, indivisible, with liberty and justice for all.
[Francis R. Bellamy, Baptist minister, *Pledge to the Flag,* 1892 (note: the words under God were subsequently added)]

America! America! God shed his grace on thee.

Founded by Christian missionaries in the mid-1700s as the Mission San Antonio de Valero, this Catholic mission would later be known as The Alamo. After English-descent settlers revolted against Mexican rule, it was the site of a "last stand" by patriotic Texans on 6 March 1836. Today it is a museum in San Antonio (photo courtesy of U.S. National War College).

O, beautiful for Patriot Dream, that sees beyond the years,
Thine alabaster cities gleam, undimmed by human tears!
America! America! God shed His grace on thee,
And crown thy good with brotherhood, from sea to shining sea.
[Katherine L. Bates, *America the Beautiful*, 1893]

Liberty means responsibilities.
[George Bernard Shaw, *Man and Superman*, 1903]

Our flag is our national ensign, pure and simple. Behold it! Listen to it! Every star has a tongue, every stripe is articulate.
[Robert C. Winthrop (1809-1894), U.S. Senator]

Liberty means responsibilities.

I therefore believe it is my duty to my country to love it, to support its Constitution, to obey its laws, to respect its flag, and to defend it against all enemies.
[William Tyler Page, *The American Creed*, 1917]

The meaning of America is not a life without toil. Freedom is not only bought with a great price, it is maintained by unremitting effort. *[and also]* Patriotism is easy to understand in America. It means looking out for yourself by looking out for your country.

[J. Calvin Coolidge Jr., U.S. President, *The Price of Freedom*, 1924]

Here lie men who loved America. . . . Out of this, and from the suffering and sorrow of those who mourn this, will come, we promise, the birth of a new freedom for the sons of men everywhere.

The only thing we have to fear is fear itself.

[Franklin D. Roosevelt, U.S. President, 4 March 1933]

It is the love of country that has lighted, and that keeps glowing, the holy fire of patriotism.

[J. Horace McFarland (1859-1948), American political observer]

We shall fight on the beaches. We shall fight in the landing grounds. We shall fight in the fields and in the streets. We shall fight in the hills. We shall never surrender.

[Sir Winston Churchill, British Prime Minister, in the House of Commons, 4 June 1940, after the debacle at Dunkirk]

Here lie men who loved America. . . . Out of this, and from the suffering and sorrow of those who mourn this, will come, we promise, the birth of a new freedom for the sons of men everywhere.

[Roland B. Gittelsohn, USN Chaplain, at a memorial service at graves of U.S. Marines killed on Iwo Jima, March 1945]

The last time any of his fellow prisoners heard from him, Captain Versace was singing *God Bless America* at the top of his voice.

America is a passionate idea – American is a human brotherhood.

[Max Lerner, *Actions and Passions*, 1949]

Only our individual faith in freedom can keep us free.

[Gen. Dwight D. Eisenhower, USA, U.S. President, 1952]

Patriotism is not short frenzied outbursts of emotion, but the tranquil and steady dedication of a lifetime.

[Adlai E. Stevensen, American diplomat, 27 August 1952]

Live free or die.

[New Hampshire state motto]

I am an American, fighting in the forces which guard my country and our way of life. I am prepared to give my life in their defense.

[Article I, Code of Conduct, U.S. Armed Forces, 1955]

Let every nation know, whether it wishes us well or ill, that we shall pay any price, bear any burden, meet any hardship, support any friend, oppose any foe, to assure the survival and success of liberty. *[and also]* Ask not what your country can do for you. Ask what you can do for your country.

[John F. Kennedy, U.S. President, 20 January 1961]

If we lose freedom here, there is no place to escape to. This is the last stand on Earth.

If we lose freedom here, there is no place to escape to. This is the last stand on Earth.

[Ronald Reagan, later U.S. President, 1964]

The last time any of his fellow prisoners heard from him, Captain Versace was singing *God Bless America* at the top of his voice.

[from Medal of Honor citation, Capt. Humbert Roque "Rocky" Versace, USA, a prisoner-of-war who was dragged out of his bamboo cage and shot by his captors, 26 September 1965]

. . . he sustained multiple fragmentation wounds from exploding grenades as he ran to an abandoned machine gun position . . . Corporal Maxam's position received a hit from a rocket propelled grenade, knocking him backwards and inflicting severe fragmentation wounds to his face and right eye. Although momentarily stunned and in intense pain, Corporal Maxam courageously resumed his firing position and subsequently was struck again by small arms fire. . . . The [enemy] threw hand grenades and directed recoilless rifle fire against him, inflicting two additional wounds. Too weak to reload his machine gun, Corporal Maxam fell to a prone position and valiantly continued to deliver effective fire with his rifle. After one and a half hours, during which he was hit repeatedly by fragments from exploding grenades and concentrated small arms fire, he succumbed to his wounds . . . . He gallantly gave his life for his country.

[Medal of Honor citation, Cpl. Larry L. Maxam, USMC, 1968]

He insisted on giving his life so that forty of his fellow Marines might live and triumph. He had freely chosen loyalty above life.

He insisted on giving his life so that forty of his fellow Marines might live and triumph. He had freely chosen loyalty above life.

[1stLt. Michael Stick, USMC, speaking of Cpl. Larry L. Maxam, USMC, killed-in-action in Vietnam on 2 February 1968]

For those who fight for it, freedom has a flavor the protected will never know.

For those who fight for it, freedom has a flavor the protected will never know.

[unidentified PFC, USMC, at Khe Sanh, Vietnam, April 1968]

Ask not what your country can do for you. Ask what you can do for your country.

We are honored to have had the opportunity to serve our country under difficult circumstances. We are profoundly grateful to our Commander-in-Chief and to our Nation for this day. God bless America!

[Capt. Jeremiah Denton Jr., USN, former POW in North Vietnam (later U.S. Senator), upon his release on 13 February 1973]

War is hell. But some things are worse than hell – slavery being one.

[Capt. Jeremiah Denton Jr., USN (later U.S. Senator), c. 1974]

Let the Fourth of July always be a reminder that here in this land, for the first time, it was decided that man is born with certain God-given rights; that government is only a convenience created and managed by the people, with no power of its own except those voluntarily granted to it by the people. We sometimes forget that great truth, and we never should.

[Ronald Reagan, U.S. President, 4 July 1981]

It is my heritage to stand erect, proud and unafraid. To think and act for myself, enjoy the benefit of my creations, to face the whole world and boldly say, "I am a free American."

[excerpt from *The Republican Creed*]

The Marines knew they were fighting for freedom, and they had an enormous respect for basic American values.

[SSgt. Arvin S. Gibson, USA, *In Search of Angels*, 1990]

I swore to defend my nation against all enemies, foreign and domestic. It doesn't get any simpler.

You can just call me an American Patriot.

[Maj. Harry R. "Bob" Mills, USMC, to a friend, 1991]

I swore to defend my nation against all enemies, foreign and domestic. It doesn't get any simpler. Stop trying to understand us.

[unidentified corporal, USMC, in *A Sense of Values*, 1994]

The red, white, and blue flag of the United States of America triumphantly fluttered in the stiff breeze atop Hill 881 North, deep in the heart of Indochina.

[Capt. Marion F. Sturkey, USMC, *Bonnie-Sue*, 1996]

The price of freedom is not cheap.

[Marine Corps Association, *Guidebook for Marines*, 2001]

America's freedom, and the values that protect us in the face of evil, are our great and glorious cause.

I gave more to America than I ever took from America, and I am proud of it. Semper Fi! And, God bless you all!

[Col. Wayne Shaw, USMC, at his retirement ceremony]

America's freedom, and the values that protect us in the face of evil, are our great and glorious cause.

[Capt. John McCain, USN, also U.S. Senator, September 2001]

I gave more to America than I ever took from America, and I am proud of it.

Semper Fi, brothers! God bless the United States and the Corps!

[2ndLt. John Fales, USMC, 12 September 2001]

This will be a battle between good and evil . . . It will be an honor to fight for God, Country, and the good of mankind.

[LCpl. Thomas Macedo, USMC, 20 September 2001]

There will never be another nation such as ours. Take good care of her. The fate of the world depends upon it.

[Capt. John McCain, USN, also U.S. Senator, at the U.S. Naval Academy, 9 October 2001]

It will be an honor to fight for God, Country, and the good of mankind.

God has blessed America with much bounty and many fine men and women through the years, who have risked their lives – then given them – to preserve our liberty.

[David Russell, American Legion chaplain, 18 October 2001]

We speak *English* – not Spanish, Lebanese, Arabic, Chinese, Japanese, Russian, or any other language.

Vocal patriotism is a form of protest against terrorism.

[C. Welton Gaddy, in *Liberty*, 2002]

The currency of freedom is blood. This land is not about perks and privilege. It is about service and sacrifice.

[Capt. Al C. Allen, USN Chaplain, 26 July 2003]

We speak *English* – not Spanish, Lebanese, Arabic, Chinese, Japanese, Russian, or any other language. Therefore, if you wish to become part of our society, learn the language!

[editorial in the *Sidney Herald* newspaper, 21 March 2007]

Pray for America and her men and women in the military, for they are our brightest and best.

[SSgt. Raymond P. "Ray" Kittles, USA, quoted in *The Lakelands Leatherneck*, 31 July 2007]

An American military veteran is a person who – at one point in his life – wrote a check payable to the United States of America for an amount "up to and including my life."

[flyers posted in the Family YMCA, Greenwood, South Carolina, for a Veterans Day program, November 2007]

How about American heroism and sacrifice? How about every Memorial Day? Every Veterans Day? Every Independence Day? Every Medal of Honor ceremony?

> [Michelle Malkin, "2 Michelles, 2 Americas, Shame vs Pride" in *National Review*, reprimanding a U.S. presidential candidate's wife, who had verbally admitted her long-term lack of pride in America, 20 February 2008]

We are all lucky, because we live in the best-est country in the whole world. God bless America!

> [A third grade student, speaking at a Veterans Day program at her school near McCormick, South Carolina, 11 November 2008]

Any listing of American patriotic statements should include lyrics from two popular songs:

Irving Berlin wrote God Bless America in 1938, based upon similar lyrics he had composed in 1918. Kate Smith popularized the song by introducing it and singing it on her radio broadcast on Armistice Day (now, *Veterans Day*) later that year. Rights to the lyrics are held by a third party.

The more recent God Bless the USA was written and introduced by Lee Greenwood in 1984. Although popular at that time, 17 years later the lyrics achieved near immortality after the terrorist attacks in New York and Washington in 2001. Rights to the lyrics are held by a third party.

Somber Reflections upon Combat

God of our fathers, known of old,
Lord of our far-flung battle line,
Beneath whose awful hand we hold
Dominion over palm and pine –
Lord God of Hosts, be with us yet,
Lest we forget – Lest we forget!
[Rudyard Kipling, *Recessional*, 1897]

Warfare and history can not be separated. In a sense, the history of mankind *is* the history of war. Those who have tasted combat know that war is obscene, terrible beyond mortal description. Yet, throughout the centuries war has remained with us.

Saint Matthew reminds us that the world will experience "wars and rumours of wars" and "nation shall rise against nation, and kingdom against kingdom." History has proven Saint Matthew to be correct. Some nations resort to warfare out of need or greed or religious fervor. Other countries take up arms in self defense, to battle terrorism and tyranny, or to thwart assorted evils.

Historically the individual soldiers, the warriors, the centurions, the legionnaires, the individual combatants, have come from the ranks of common men. Bound together by unity of cause and dedication to their brothers-in-arms, they are revered by their countrymen. Yet, warfare rarely brings fame and glory. Instead, warfare brings misery, privation, horror, and indiscriminate death. War's toll is cruelty, brutality, and sheer madness.

> Warfare rarely brings fame and glory. Instead, warfare usually brings misery, privation, horror, and indiscriminate death.

> Those who have tasted combat know that war is obscene, terrible beyond mortal description.

The following words, arranged in chronological order, come from men who have seen the face of war. Fame and glory are fleeting. Combat is ugly and obscene. For the individual warrior, often the only victory that remains is eternal loyalty to his fellow brothers-in-arms. They become his friends for life:

War is sweet to those who have never experienced it.
  [Pindar (522-443 BC), Greek poet]

Go tell the Spartans, thou that passeth by,
That here, obedient to the laws, we lie.
  [epitaph for Spartan soldiers who fell in battle while holding the pass at Thermopylae, 480 BC]

Only the dead have seen the end of war.
  [Plato (428-347 BC), Greek philosopher]

Only the dead have seen the end of war.

I did not mean to be killed today.
  [Vicomte de Turenne, a wounded French soldier, as he lay dying after the Battle of Salzbach, 1675]

These are the times that try men's souls. The summer soldier and the sunshine patriot will, in this crisis, shrink from the service of their country. . . . Tyranny, like Hell, is not easily conquered.
  [Thomas Paine, *The American Crisis*, 1776]

War is a rough, violent trade.
  [Johann C. Schiller, *The Piccolomini*, 1799]

Thank God I have done my duty.
  [final words of Viscount Horatio Nelson, RN, mortally wounded aboard his *HMS Victory* off Cape Trafalgar, 21 October 1805]

Tyranny, like Hell, is not easily conquered.

There is no uglier spectacle than two men with clenched teeth and hellfire eyes, hacking at one another's flesh, converting precious living bodies, and priceless human souls, into nameless putrescence.
  [Thomas Carlyle, *Past and Present*, 1843]

Thousands of white crosses in a U.S. military cemetery near Hamm, Luxembourg, mark graves of U.S. servicemen killed during World War II (photo by the author, Marion Sturkey).

On fame's eternal camping ground
Their silent tents are spread,
And glory guards with solemn round
The bivouac of the dead.

[Theodore O'Hara, *The Bivouac of the Dead*, 1847 (by an Act of the U.S. Congress, displayed in every U.S. National Cemetery)]

Theirs not to make reply,
Theirs not to reason why,
Theirs but to do and die.
Into the valley of Death rode the six hundred.

[Alfred Tennyson, *The Charge of the Light Brigade*, 1854]

If a man had told me twelve months ago that men could stand such hardships, I would have called him a fool.

[Lt. James H. Langhorne, CSA, 8 January 1862]

Why does Colonel Grigsby refer to me to learn how to deal with mutineers? He should shoot them where they stand.

[LtGen. Thomas J. "Stonewall" Jackson, CSA, in regard to requests for discharge from 12-month volunteers, May 1862]

War is a rough, violent trade.

No tongue can tell, no mind can conceive, no pen can portray, the horrible sights I witnessed this morning.

[Capt. John Taggart, USA, South Mountain, 17 September 1862]

Remember that the enemy you engage have no feelings of kindness or mercy towards you.

[MGen. Thomas C. Hindman, CSA, 7 December 1862]

It is well that war is so terrible, for we would grow too fond of it.

[Gen. Robert E. Lee, CSA, Fredericksburg, 13 December 1862]

War is cruelty and you cannot refine it.

Major, tell my father I died with my face to the enemy.

[final words of Col. Issac E. Avery, CSA, spoken to his adjutant as he lay mortally wounded at Gettysburg, 2 July 1863]

Well, general, let's bury these poor men and say no more about it.

[Gen. Robert E. Lee, CSA, to MGen. A.P. Hill, CSA, in regard to the Confederate dead at Briscoe Station, 14 October 1863]

We cannot dedicate, we cannot consecrate, we cannot hallow this ground. The brave men, living and dead, who struggled here have consecrated it far above our poor power to add or detract. The world will little note nor long remember what we say here, but it can never forget what they did here.

[Abraham Lincoln, U.S. President, dedicating the U.S. National Cemetery at Gettysburg battlefield, 19 November 1863]

No tongue can tell, no mind can conceive, no pen can portray, the horrible sights I witnessed this morning.

Don't worry, they couldn't hit an elephant at this dis . . . .

[final interrupted words of Gen. John Sedgwick, USA, shot and killed by a Confederate sniper at Spotsylvania, 8 May 1864]

War is cruelty and you cannot refine it.
[Gen. William T. Sherman, USA, in a letter, 2 September 1864]

There is many a boy here today who looks on war as all glory. But, boys, war is Hell!

None can realize the horrors of war, save those actually engaged. The dead lying all about, unburied to the last. My God! My God! What a scourge is war!
[Samuel Johnson, Sixth Georgia, CSA, in a letter, 1864]

When I was taken prisoner I weighed 165 pounds, and when I came out I weighed 96 pounds and was considered stout compared to many I saw there.
[Pvt. A.S. Clyne, USA, former POW at Andersonville, 1865]

There's only one truth about war – people die.
[Gen. Philip H. Sheridan, USA (1831-1888)]

Not for fame or reward, not for place or rank,
Not lured by ambition or goaded by necessity;
But in simple obedience to duty as they understood it,
These men suffered all, sacrificed all,
Dared all – and died.
[eulogy by Rev. Randolph H. McKim, CSA chaplain, inscribed on the Confederate Memorial in Arlington National Cemetery]

Don't cheer, men. The poor devils are dying.

There is many a boy here today who looks on war as all glory. But, boys, war is Hell! . . . It is only those who have neither fired a shot nor heard the shrieks and groans of the wounded who cry aloud for blood, more vengeance, more destruction.
[Gen. William T. Sherman, USA, speaking to military veterans and young men, 12 August 1880]

War loses a great deal of its romance after a soldier has seen his first battle.
[Col. John Mosby, CSA, *Mosby's War Reminiscences*, 1887]

Don't cheer, men. The poor devils are dying.

[Capt. John Philip, USN, as his *USS Texas* passed the burning Spanish warship *Vizcaya* at Santiago, 3 July 1898]

If I come out of this war alive, I will have more luck than brains.

[Baron Capt. Manfred von Richthofen (*The Red Baron*), German Flying Service, in a letter to his mother, 1914]

I have seen war, and faced modern artillery, and I know what an outrage it is against simple men.

[Thomas M. Kettle, *The Ways of War,* 1915]

The effects of the successful gas attack were horrible. I am not pleased with the idea of poisoning men. Of course the entire world will rage about it at first – and then imitate us.

[Rudolph Binding, *A Fatalist at War*, in regard to the German use of poison gas at Vijfwege, Belgium, in April 1915]

With a bullet through his head, he fell from an altitude of 9000 feet, a beautiful death.

[Baron Capt. Manfred von Richthofen (*The Red Baron*), German Flying Service, describing the death of his friend near Verdun, France, 1 May 1916]

My God! Did we really send men to fight in that?

My God! Did we really send men to fight in that?

[LtGen. Sir Launcelot E. Kiggell, British Army, upon seeing the mud and carnage after the Battle of Passchendaele, 1917]

What's the matter? Do you think that perhaps I will not return?

[Baron Capt. Manfred von Richthofen (*The Red Baron*), German Flying Service, before his final flight, 21 April 1918]

On 30 September 1966, infantrymen from the Fifth Marines are locked in battle with North Vietnamese soldiers near the Vietnamese DMZ (photo courtesy of U.S. Marine Corps).

I have seen blood running from the wounded. I have seen men coughing out their gassed lungs. I have seen the dead in the mud. I have seen two hundred limping, exhausted men come out of the line, the survivors of a regiment of one thousand that went forward forty-eight hours before.

[Franklin D. Roosevelt, Assistant Secretary of the Navy (later U.S. President), at Belleau Wood, France, June 1918]

I have only two men out of my company and 20 out of some other company. We need support, but it is almost suicide to try to get it here as we are swept by machine gun fire and a constant barrage is on us. I have no one on my left and only a few on my right. I will hold.

I have seen blood running from the wounded. I have seen men coughing out their gassed lungs. I have seen the dead in the mud.

[1stLt. Clifton B. Cates, USMC, in France, 19 July 1918]

War would end if the dead could return.

[Stanley Baldwin (1867-1947), British Prime Minister]

War, like any other racket, pays high dividends to the very few.
[MGen. Smedley D. Butler, USMC, 1933]

Enemy on island. Issue in doubt.
[last American radio message, Wake Island, 23 December 1941]

Every day kill just one, rather than today five, tomorrow ten. Then your nerves are calm and you can sleep good. You have your drink in the evening, and the next morning you are fit again.
[Col. Erich Hartmann, Luftwaffe, 352 kills, World War II]

Older men declare war. But it is youth that must fight and die.
[Herbert Hoover, former U.S. President, 27 June 1944]

Older men declare war. But it is youth that must fight and die.

The beach was a sheet of flame backed by a wall of black smoke, as though the island was on fire. . . . We piled out of our Amtrac amid blue-white Japanese machine gun tracers and raced inland.
[PFC Eugene B. Sledge, USMC, Peleliu, 15 September 1944]

Such a sight on that beach! Wrecked boats, bogged down jeeps, tanks burning, casualties scattered all over!
[Michael Kelecher, USN surgeon, Iwo Jima, February 1945]

All I wanted to get out of Iwo Jima was my fanny and dog tags.
[Cpl. Edward Hartman, USMC, Iwo Jima, March 1945]

You never knew when you were drawing your last breath. You lived in total uncertainty, on the brink of the abyss, day after day.

How can I feel like a hero, when I hit the beach with two-hundred-and-fifty buddies, and only twenty-seven of us walked off alive?
[PFC Ira A. Hayes, USMC, on 16 April 1945 after being called a "hero" for his role in raising the American flag on Iwo Jima]

You never knew when you were drawing your last breath. You lived in total uncertainty, on the brink of the abyss, day after day.

[PFC Eugene B. Sledge, USMC, Okinawa, 1945]

Another improvement was that we built our gas chambers to accommodate two thousand people at one time.

[Rudolf Hess, former Deputy Fuhrer of Germany, during his imprisonment in England after World War II]

A million deaths is a mere statistic.

A million deaths is a mere statistic.

[Josef W. Stalin (1879-1953), Russian dictator]

Consider yourselves already dead. Once you accept that idea, it won't be so tough.

[actor Gregory Peck, in the movie, *Twelve O'Clock High*, 1949]

The experience helped me realize how fragile life is. There could be two of you standing there – and in the next minute, only one.

[Pvt. Jack McCorkle, USMC, speaking of the fighting at Chosin Reservoir, Korea, in December 1950]

The staff intelligence officer handed me the pre-strike photos, the coordinates of the target, and told me to get on with it. He didn't mention that the bridges were defended by 56 radar-controlled anti-aircraft guns.

[Capt. Paul N. Gray, USN, speaking of attacking the crucial bridges at Toko-ri, Korea, on 12 December 1951]

There could be two of you standing there – and in the next minute, only one.

The situation of the wounded is particularly tragic. They are piled on top of each other in holes that are completely filled with mud and devoid of any hygiene.

[Bernard B. Fall, *Hell in a Very Small Place*, 1966, quoting a French Army radio message from Dien Bien Phu on 5 May 1954]

The Legionnaire next to me disintegrated. Nothing was left of him except little pieces of raw meat. Death was spitting all around us. Men were falling like flies.

[a survivor of Dien Bien Phu in 1954, speaking years later]

> Consider yourselves already dead. Once you accept that idea, it won't be so tough.

The wounded were still lying there just like on the first day, intermingled with men who had died several days ago and were beginning to rot. They were lying there unattended in the tropical sun, being eaten alive by the rats and vultures. If only they had all been dead! *[and also]* As night fell over Dien Bien Phu the Legionnaires fixed bayonets in the ghostly light of the parachute flares and – 600 against 40,000 – walked into death.

[Bernard B. Fall, *Street Without Joy*, 1961]

The survivors would envy the dead.

[Nikita Khrushchev, Premier of the Soviet Union, speaking of the possibility of global nuclear war, 1962]

> The wounded were still lying there, just like on the first day, intermingled with men who had died several days ago and were beginning to rot. They were . . . being eaten alive by the rats and vultures. If only they had all been dead!

By that time every Marine had been wounded. The living took the ammunition of the dead and lay under a moonless sky, wondering about the next assault.

[Capt. Francis J. West Jr., USMC, in *Small Unit Action in Vietnam, Summer 1966*, 1967]

We were being attacked by a thousand men. We just couldn't kill them fast enough.

[Sgt. John J. McGinty, USMC, quoted in *U.S. Marines in Vietnam, an Expanding War, 1966*, 1982]

One [survivor was] about eighteen, covered with gunpowder and dirt, black under the eyes. They were glassy. He was exhausted. Man, he looked bad, real bad. He said he had his bayonet fixed all night. I asked him if he had been scared, and he said, "Yeah." Right before daylight he had one bullet left. One bullet, just one bullet. So he started throwing rocks at the [enemy] in the dark. You know, tryin' to make 'em think the rocks were grenades. Only one bullet left. He was saving it for the final charge. He told me he realized he was gonna' die. Then, once he accepted that, he wasn't scared anymore.

He told me he realized he was gonna' die. Then, once he accepted that, he wasn't scared anymore.

[Capt. William T. Holmes, USMC, referencing a conversation in Vietnam on 9 August 1966 (quoted in *Bonnie-Sue*, 1996)]

Air strikes coming every 30 seconds. The ground trembles continuously. Once again I feel the end is near – at least for me.

He flew to Phu Bai, took off for Marble Mountain, and was never heard from again.

[Maj. Harry R. "Bob" Mills, USMC, speaking years later about a friend MIA in Vietnam on 6 October 1966]

He charged a machinegun with hand grenades, trying to save some guys. He didn't have to do it. He got slaughtered. He had two kids.

[Capt. Otto H. Fritz, USMC, speaking about a friend in Vietnam in 1966 (quoted in *Bonnie-Sue*, 1996)]

Two machineguns keep up intense fire. NVA now have us almost surrounded. I have a terrible feeling I will never see my family again. . . . Air strikes coming every 30 seconds. The ground trembles continuously. Once again I feel the end is near – at least for me. I get an uncontrollable case of the shakes. I wonder if I ever had what it takes to be a Marine and conclude that I never did and don't now.

You go ahead, I'm a dead man.

[Arnaud de Borchgrave, combat correspondent, "The Battle for Hill 400" in *Newsweek Magazine*, 10 October 1966]

Charred burned-out wreckage is all that remains of a U.S. helicopter downed at Khe Sanh, South Vietnam, on 27 January 1967 (photo by the author, Marion Sturkey).

Otto, you go ahead, I'm a dead man.

[final words of 1stLt. Steve Sayer, USMC, moments before he was killed in Vietnam, 10 December 1966]

I put my hand down, counted my fingers with my thumb, and then went back to shooting. I was afraid to look at my arm.

[GySgt. Gareth L. "Red" Logan, USMC, writing years later about combat near Khe Sanh, Vietnam, on 25 April 1967]

Mike Company ceased to exist on that day. Out of 190 men, only 26 were left standing.

[Austin Deuel, in *Vietnam Magazine*, describing 30 April 1967]

Where are those [expletive] choppers? All my emergency medevacs are dead! All my priorities are now emergencies!

[HM3 Thomas Lindenmeyer, USN, Vietnam, 2 July 1967]

He was burning to death in the plane and couldn't get out. He was [screaming for] someone to tell his wife that he loved her, and for someone to shoot him.

I don't think I'll be talking to you again. We're being overrun.

[final radio message from Capt. Warren O. Keneipp, USMC, killed in Vietnam (staked out and decapitated), 2 July 1967]

He was blown in half for a [expletive] place that had no strategic value, no military value, no sense to it, save to prove to Russia or China or North Vietnam or God or somebody that nineteen year old low- and middle-class Americans would die for their country.

[a teenage U.S. Marine at Con Thien, Vietnam, 1967]

He was burning to death in the plane and couldn't get out. He was [screaming for] someone to tell his wife that he loved her, and for someone to shoot him.

[Chaplain Ray Stubbe, USN, *The Final Formation*, 1995, quoting a witness to the death of a pilot at Khe Sanh on 23 August 1967]

Out of 190 men, only 26 were left standing.

The lead aircraft disintegrated in the air. Two pilots, two crewmen, and one passenger were on board. There were no survivors.

[*HMM-262 Command Chronology*, 31 August 1967]

We found part of Scribner's helmet with part of his head still inside of it. . . . [It] looked like the rocket went off right in his lap. There was nothing left.

[Chaplain Ray Stubbe, USN, *The Final Formation*, 1995, in regard to the death of a Marine at Khe Sanh on 24 January 1968]

Sometimes in the morning we'd see three- or four-hundred bodies out along the wire.

We huddled together in the bunker, shoulders high and necks pulled in to leave no space between helmet and flak jacket. There is no describing an artillery barrage. The earth shakes, clods of dirt fall from the ceiling, and shrapnel makes a repulsive [noise] singing through the air.

[John Donnelly, combat correspondent, "Drawing the Noose" in *Newsweek Magazine*, 5 February 1968]

Everything I see is blown through with smoke, everything is on fire everywhere. It doesn't matter that memory distorts; every image, every sound comes back out of smoke and the smell of things burning. *[and also]* The Grunts themselves knew: the madness, bitterness, the horror and doom of it. *[and also]* The belief that one Marine was better than ten Slopes saw Marine squads fed in against known NVA platoons, platoons against companies, and on and on, until whole battalions found themselves pinned down and cut off. That belief was undying, but the Grunt was not.

The Grunts themselves knew: the madness, bitterness, the horror and doom of it.

[Michael Herr, *Dispatches*, 1968]

Sometimes in the morning we'd see three- or four-hundred bodies out along the wire. *[and also]* Every inch of the runway was zeroed in, and if an airplane tried to land they just walked artillery rounds right up the centerline. *[and also]* They had the glideslope zeroed-in with .50 caliber machineguns and they knew exactly where to shoot to hit you on it. They'd just listen for you and start laying fire down the glideslope. If you were on it, you were drilled.

[LtCol. David L. Althoff, USMC, "Helicopter Operations at Khe Sanh" in *Marine Corps Gazette*, May 1969]

It was raining. I was a replacement for a company commander who had been killed the night before. The tank lurched to a halt. I jumped off, walked over to a hole and asked, "Where is the CP?" A filthy, soaking wet Marine continued bailing out his hole with a C-ration can and answered, "You're in it." I asked for the battalion commander. He answered, "You're looking at him."

[Maj. M.P. Caulfield, USMC, "India Six" in *Marine Corps Gazette*, July 1969]

Jerry, I'm in bad shape. They are giving me almost nothing to eat. I'm down to a hundred pounds, and I haven't crapped in twenty-six days. I don't remember how long I've been in irons, but it's been weeks. I don't know if I can make it.

War runs best on evil. How else can you convince boys to kill one another day after day?

[RAdm. Jeremiah A. Denton Jr., USN (later U.S. Senator), paraphrasing a fellow POW in *When Hell Was in Session*, 1976]

I promised God that I would quit smoking and I would never touch a whore, not even get a hand job, and I would believe in Him if He would only let me live.

Samuels looked down and saw that his left leg was flipped crazily to one side midway down below the knee. There was no way his leg could be lying there like that and still be – still be attached.

[C.D.B. Bryan, *Friendly Fire*, 1976]

War runs best on evil. . . . How else can you convince boys to kill one another day after day? *[and also]* War is not killing. Killing is the easiest part of the whole thing. Sweating twenty-four hours a day, seeing guys drop all around you of heatstroke, not having food, not having water, sleeping only three hours a night for weeks at a time, that's what war is.

[Mark Baker, *NAM*, 1981]

It was a slaughter. No better than lining people up on the edge of a ditch and shooting them in the back of the head. I was doing it enthusiastically.

There is joy, true joy, in being alive when so many around you are not.

[unidentified helicopter gunner, quoted in *NAM*, 1981]

I couldn't get over how bizarre it was. We would decide to stop killing each other for a few days, and then start again. *[and also]* I felt like a worm on a string. The tracers rushed past us like a line of UFOs in a hurry. I promised God that I would quit smoking and I would never touch a whore, not even get a hand job, and I would believe in Him if He would only let me live.

The tracers rushed past us like a line of UFOs in a hurry.

[WO Robert Mason, USA, *Chickenhawk*, 1984]

The enduring emotion of war, when everything else has faded, is comradeship. A comrade in war is a man you can trust with anything, because you have trusted him with your life. *[and also]* In war the line between life and death is gossamer thin; there is joy, true joy, in being alive when so many around you are not.

Death is so commonplace it doesn't shock you anymore.

[William Broyles Jr., "Why Men Love War" in *Esquire*, 1984]

The POWs I saw were very thin; they were covered with scabies – there was just skin and bones left on them. They could hardly walk, yet they were forced to carry wood from the forests. They often fell down. They were beaten by the guards.

[a South Vietnamese soldier, quoted in *Life on the Line*, 1988]

The only thing clean borne of this life is cruelty and filth.

[unidentified British private, quoted in *Eye Deep in Hell: Trench Warfare in World War I*, 1989]

I now know why men who have been to war yearn to reunite. Not to tell war stories or look at old pictures. Not to laugh or weep. Comrades gather because they long to be with men who once acted their best, men who suffered and sacrificed, who were stripped raw, right down to their humanity. . . I have never given anyone such trust. They were willing to guard

In war the line between life and death is gossamer thin.

Infantrymen and corpsmen from the Fifth Marines carry a fallen comrade toward a waiting CH-46 helicopter in Iraq on 6 April 2007 (photo courtesy of U.S. Marine Corps).

something more precious than my life. They would have carried my reputation, the memory of me. It was part of the bargain we all made, the reason we were so willing to die for one another.

[Michael Norman, USMC, *These Good Men*, 1990]

Thirty, forty, maybe fifty Marines lay twisted along both sides of the road, clumped atop each other in spots, their weapons and gear strewn down the middle of the road . . . a slaughterhouse.

[William K. Nolan, *Operation Buffalo*, 1991]

Death is so commonplace it doesn't shock you anymore. *[and also]* Flying in the night rain with fog was a death warrant.

[LtCol. H. Lee Bell, USMC, *1369*, 1992]

He was just a kid, as was I. He confided to me that he had never even kissed a girl before. . . . Unfortunately, he never got the chance. I think his mother would be happy to know that only God and her [sic] ever knew the tenderness of his kiss.

[Chaplain Ray Stubbe, USN, *The Final Formation*, 1995, quoting a friend of PFC Bruce Cunningham, USMC, KIA in Vietnam]

Our government does not want America to know that our darkest secret is that we killed many Americans in cold blood. They were tortured to death in prison, or simply killed outright from fear they would try to escape. And our leaders are afraid to admit this. They were tortured to death in Hanoi.

[LtCol. Nguyen Van Thi, Vietnamese Army, quoted in *Inside Hanoi's Secret Archives: Solving the MIA Mystery*, 1995]

We did unspeakable things we won't now admit – not even to ourselves.

We did unspeakable things we won't now admit – not even to ourselves. We held the God-like power of life and death in our hands. After the war we suppressed the killer instinct, usually hiding it behind a low-key facade of casual humor. But the evil still lives in us, lurking somewhere just beneath the surface.

[Capt. Marion F. Sturkey, USMC, in his handwritten notes for "Requiem" in *Bonnie-Sue,* 1996]

War is a cruel game, a brutal game, a deadly game. *[and also]* The overwhelming sensation was that of deafening noise and bedlam. No one could hear the individual weapons firing, the bombs exploding, the shouts and screams. There was only a continuous cacophony, a horrible roar. *[and also]* Low on ammunition, the two wounded Marines crawled among their fallen friends and stripped them of all their remaining grenades and M-16 magazines. . . . The alternate radio operator, although unconscious, was still alive. Pulling him between them, they waited, for there was nothing else they could do.

No one could hear the individual weapons firing, the bombs exploding, the shouts and screams. There was only a continuous cacophony, a horrible roar.

[Capt. Marion F. Sturkey, USMC, *Bonnie-Sue,* 1996]

Men caught in the killing zone became instant dogmeat. *[and also]* Helicopter crews who survived an entire tour unscathed led charmed lives. Enemy gunfire downed 1,777 helicopters during the first five years of the war; others returned to base shot to splinters.

[Col. Joseph H. Alexander, USMC, *A Fellowship of Valor*, 1997]

Sweat gathered inside our rubber boots, and when we pulled off a sock, frozen skin came with it. Sleep was out of the question. Training enabled us to keep fighting. Surrounded as we were, there was no rear, no front, no flank.

Second place was a body-bag.

[Sgt. Werner "Ronnie" Reininger, USMC, describing fighting at Hagaru-ri, Korea, in *Leatherneck Magazine*, January 2001]

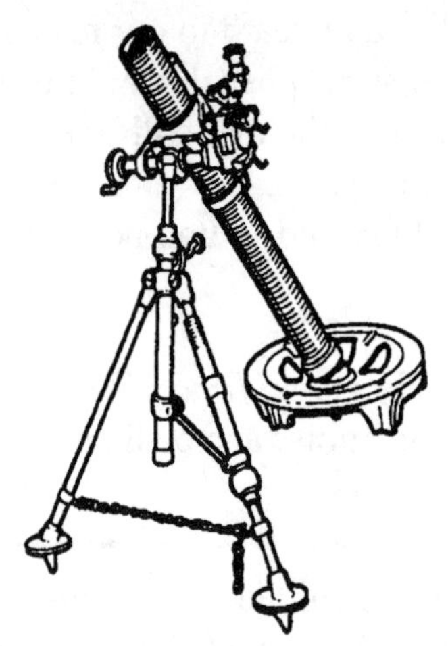

Second place was a body bag.

[Capt. Roger A. Herman, USMC, in *The Log Book*, 2001]

I watched Marines die face down in the mud protecting freedom.

[Col. Oliver North, USMC, 21 September 2001]

Beware of war-hawks who never served in the military.

[James Bradford, in *USA Today*, 17 September 2002]

I came across the body of the old man with the cane. He had a massive wound in the back of his head. He died on his back, looking at the sky, and his body was covered with flies.

[Peter Maas, in *New York Times Magazine*, 20 April 2003]

Beware of war-hawks who never served in the military.

Men caught in the killing zone became instant dogmeat.

No same person who has ever been to war wants to go to another.

[Col. Oliver North, USMC, 29 May 2003]

This is war, and in war people die, both combatants and civilians.
[Sgt. Phil P. Lucas, USMC, in *The Lakelands Leatherneck*, December 2004]

We had expected beheadings, of course, but never cannibalism! What kind of men were these?
[Col. Dick Camp, USMC, quoting a U.S. war crimes investigator (who learned Imperial Japanese Army soldiers had killed and <u>eaten</u> many American POWs during World War II) in *Leatherneck Magazine*, July 2008]

Lo, all our pomp of yesterday
Is one with Nineveh and Tyre!
Judge of the Nations, spare us yet,
Lest we forget – Lest we forget!
[Rudyard Kipling, *Recessional*, 1897]

That Flag Stands for Freedom

Three days before Christmas in 2001 a bright silver Boeing 767 airliner took off from Paris, France, and arced out over the Atlantic Ocean. With 197 passengers and crew aboard, American Airlines Flight 63 headed for Miami, Florida.

One of the passengers was a chronic societal misfit and drifter, Richard C. Reid (also known as Abdul Raheem). He was a British citizen, born in London in 1973, the son of an English mother and a Jamaican father. Reid had served time in numerous British jails for various street crimes. During his incarceration he had converted to the Islamic faith. Before he boarded the plane, Reid had packed plastic explosives and a detonator into his high-top hiking boots. In the Boeing airliner cabin he sat alone by a window, with empty seats beside him.

> Reid had secretly packed plastic explosives and a detonator into his high-top hiking boots.

Midway into the flight, high over the Atlantic, Reid began striking matches – he lit *six* of them – in an effort to light a fuse that extended into his boots (it would later be found that he had been unable to light the fuse because it was soaked with his perspiration). Two flight attendants saw what Reid was trying to do and tried to stop him. The heavily-built Reid beat them with his fists and slammed them to the floor. Five passengers joined the fight, and they finally subdued Reid. Although he screamed, kicked, and struggled, they managed to hog-tie him with several belts, headphone cords, and plastic handcuffs. The pilots diverted to Boston, Massachusetts, where Reid was officially arrested and rushed to the local jail.

> Five passengers joined the fight, and they finally subdued Reid.

In federal district court in Boston on 30 January 2003, Reid pled guilty to multiple charges including (1) attempted murder and (2) attempted use of a weapon of mass destruction. Reid called himself a "soldier." He said he was "at war" with the United States and that his attempt to destroy the plane, passengers, and crew was "an act of war." Reid told the judge: "I think I ought not to apologize for my actions. I am at war with your country."

"I ought not to apologize for my actions. I am at war with your country."

The judge, the honorable William G. Young, sentenced Reid to prison for the remainder of his natural life. The judge explained the sentence to Reid in the following manner:

– Sentencing of Richard C. Reid –

Richard C. Reid, harken now to the sentence this Court imposes upon you. On counts one, five, and six this Court sentences you to life in prison in the custody of . . . *[the judge explains the sentence for each of the eight criminal charges against Reid].*

This Court sentences you to life in prison.

Now, let me explain this to you. We are not afraid of you or any of your terrorist co-conspirators, Mr. Reid. We are Americans, and we have been through the fire before.

. . . You are not an enemy combatant. You are a terrorist. You are not a soldier in any war. You are a terrorist. . . . We do not negotiate with terrorists. We do not meet with terrorists. We do not sign documents with terrorists. We hunt them down, one by one, and bring them to justice. So, this war talk is way out of line in this Court. You're no warrior. You are a terrorist, a species of criminal that is guilty of multiple attempted murders.

You are not an enemy combatant. You are a terrorist. You are not a soldier in any war. You are a terrorist.

. . . What your able counsel, and what the equally able United States Attorneys have grappled with, and what I have, as honestly as I know how, tried to grapple with, is why you did something so horrific. What was it that led you here to this courtroom today? I have listened respectfully to what you have to say. I ask you to search your heart and ask yourself – what kind of unfathomable hate led you to do what you are guilty, and admit you are guilty, of doing?

> What was it that led you here to this courtroom today?

. . . It seems to me you hate the one thing that, to us, is most precious. You hate our freedom, our individual freedom, our freedom to live as we choose, to come and go as we choose, to believe or not believe as we individually choose. Here in this society the very winds carry freedom. They carry it everywhere, from sea to shining sea.

It is because we prize individual freedom so much that you are here in this courtroom, so that everyone can see, truly see, that justice is administered fairly, individually, and discretely. It is for freedom's sake that your lawyers are striving vigorously on your behalf . . . .

We Americans are all about freedom, because we all know that the way we treat you, Mr. Reid, is a measure of our own liberties. We will bear any burden, pay any price, to preserve our freedoms.

> You hate the one thing that, to us, is most precious. You hate our freedom.

Look around this courtroom. Mark it well. The world is not going to long remember what you or I say here. But, see that flag, Mr. Reid? That's the flag of the United States of America. That flag will fly there

> That's the flag of the United States of America. . . . That flag stands for freedom!

long after [you are] forgotten. That flag stands for freedom! And, it always will!

Mr. Custody Officer, stand him down!

<u>Postscript</u>: After his sentencing, Richard Reid was incarcerated in ADX Florence. This federal "supermax" penitentiary was built in 1983. It is reserved for the most dangerous prisoners who require the most stringent control. Usually called the "Alcatraz of the Rockies," this supermax prison is located adjacent to Highway 67 near Florence, Colorado.

– The American's Creed –

I believe in the United States of America as a government of the people, by the people, for the people; whose just powers are derived from the consent of the governed; a democracy in a republic; a Sovereign Nation of many Sovereign States; a perfect union, one and inseparable; established upon those principles of freedom, equality, justice, and humanity for which American Patriots sacrificed their lives and fortunes.

I therefore believe it is my duty to my country to love it, to support its constitution, to obey its laws, to respect its flag, and to defend it against all enemies.

[William Tyler Page, 1917, adopted by the U.S. House of Representatives, 1918]

Codes to Keep and a Poem to Remember

– Code of Conduct –

During the Korean War in the early 1950s the Chinese Army and North Korean Army captured hundreds of Americans. The prisoners then faced a deadly and unexpected new enemy, the Eastern World POW environment. Brutal torture, random genocide, lack of food, absence of medical aid, and subhuman treatment became the daily norm. Many prisoners found themselves unprepared for this *new battlefield.*

Many prisoners found themselves unprepared for this *new battlefield.*

After the war ended the U.S. Department of Defense developed an ethical guide for warriors of the U.S. Armed Forces. Officially it is the "Code of the U.S. Fighting Force." However, it is commonly called the ***Code of Conduct***. Dwight D. Eisenhower, U.S. President, approved the code in 1955. The six articles of the code embrace (1) general statements of dedication to America and the cause of freedom, (2) personal conduct on the battlefield, and (3) personal conduct as a prisoner of war.

The Code of Conduct is not part of the Uniform Code of Military Justice (UCMJ). As military doctrine, as opposed to military law, it has no punitive provisions. In the manner of the Hippocratic Oath, the Ten Commandments, and the Ethic of Reciprocity (often called *The Golden Rule* in Christian cultures), the Code of Conduct is a personal conduct mandate for American Warriors in time of war:

– Code of Conduct –

Article I: I am an American, fighting in the forces which guard my country and our way of life. I am prepared to give my life in their defense.

Article II: I will never surrender of my own free will. If in command, I will never surrender the members of my command while they still have the means to resist.

Article III: If I am captured I will continue to resist by all means available. I will make every effort to escape and [to] aid others to escape. I will accept neither parole nor special favors from the enemy.

Article IV: If I become a prisoner of war, I will keep faith with my fellow prisoners. I will give no information nor take part in any action which might be harmful to my comrades. If I am senior, I will take command. If not, I will obey the lawful orders of those appointed over me and back them up in every way.

Article V: When questioned, should I become a prisoner of war, I am required to give name, rank, service number, and date of birth. I will evade answering further questions to the utmost of my ability. I will make no oral or written statements disloyal to my country and its allies or harmful to their cause.

Article VI: I will never forget that I am an American, fighting for freedom, responsible for my actions, and dedicated to the principles which made my country free. I will trust in my God and in the United States of America.

– General Orders –

Guard duty is nothing new. The legions of Julius Caesar and the hordes of Genghis Kahn utilized sentries, often called sentinels, to guard their camps day and night. Without diligent and alert sentries an army might never get the chance to march against its foe. During the Civil War in America in the 1860s, sentries were the *watchmen* for both Union and Confederate armies. Guard duty then, as now, was serious business. In fact, the CSA Articles of War stated:

> Any sentinel who shall be found sleeping upon his post, or who shall leave his post before being regularly relieved, shall suffer death, or other such punishment as may be inflicted by the sentence of a court martial.

Today, ***General Orders*** for the U.S. Armed Forces are bedrock upon which sentries enforce military security in the United States and throughout the world. General Orders dictate the conduct of sentries on guard duty. These orders apply at all bases and outposts in time of peace, and in time of war.

Recruits and officer candidates must memorize these General Orders. Woe be unto the unfortunate trainee who can not shout out, verbatim and without hesitation, all of them, for his failure will prompt a firestorm of wrath from his superiors. There is sound logic for this rigid training. General Orders will guide warriors on guard duty throughout their years in the U.S. Armed Forces:

– United States Army –

The U.S. Army keeps directives for its sentries general in nature, preferring to rely upon Special Orders for necessary details and instructions. There are three General Orders for Army soldiers:

General Order No. 1: I will guard everything within the limits of my post and quit my post only when properly relieved.

General Order No. 2: I will obey my Special Orders and perform all of my duties in a military manner.

General Order No. 3: I will report violations of my Special Orders, emergencies, and anything not covered in my instructions to the Commander of the Relief.

– United States Navy –

The U.S. Navy has given its sentries 11 General Orders. They are almost identical to those given to Marine Corps sentries. General Orders for Navy sailors are listed below:

General Order No. 1: To take charge of this post and all government property in view.

General Order No. 2: To walk my post in a military manner, keeping always on the alert and observing everything that takes place within sight or hearing.

General Order No. 3: To report all violations of orders I am instructed to enforce.

General Order No. 4: To repeat all calls from posts more distant from the guardhouse than my own.

General Order No. 5: To quit my post only when properly relieved.

General Order No. 6: To receive, obey, and pass on to the sentry who relieves me all orders from the commanding officer, command duty officer, officer of the deck, and officers and petty officers of the watch only.

General Order No. 7: To talk to no one except in line of duty.

> To talk to no one except in line of duty.

General Order No. 8: To give the alarm in case of fire or disorder.

General Order No. 9: To call the officer of the deck in any case not covered by instructions.

General Order No. 10: To salute all officers and all colors and standards not cased.

General Order No. 11: To be especially watchful at night and, during the time for challenging, to challenge all persons on or near my post and to allow no one to pass without proper authority.

– United States Marine Corps –

The origin of General Orders for U.S. Marines on guard duty has been lost in the fog of history. The Marine Corps is part of the Department of the Navy, and the 11 General Orders for Marine sentries are almost identical to those for Navy sentries. Perhaps Marines originated the General Orders and the Navy later adopted them. Or, perhaps it was the other way around. In either event the only difference between General Orders for sailors and General Orders for Marines is a minor wording nuance due to different Navy and Marine terminology. Like the Navy, the Marine Corps has 11 General Orders. The slightly different wording of the Marines' General Order No. 6 and General Order No. 9 follows:

The origin of General Orders for U.S. Marines on guard duty has been lost in the fog of history.

General Order No. 6: To receive, obey, and pass on to the sentry who relieves me all orders from the commanding officer, officer of the day, and officers and noncommissioned officers of the guard only.

General Order No. 9: To call the corporal of the guard in any case not covered by instructions.

– United States Air Force –

The U.S. Air Force uses specialized security units on its bases. First called *Security Police* and later *Air Police*, the current *Air Force Security Forces* provide guard duty functions, access control, law enforcement, and military security. These Air Force security forces have three basic General Orders:

> **General Order No. 1:** I will take charge of my post and protect all personnel and property for which I am assigned until properly relieved.
>
> **General Order No. 2:** I will report all violations of instructions which I am required to enforce and contact my supervisor in those cases not covered by instruction.
>
> **General Order No. 3:** I will sound the alarm in case of disturbance or emergency.

– In Flanders Fields –

My God! Did we really send men to fight in that?
[LtGen. Sir Launcelot Kiggell, British Army, 1917]

By 1915 in Europe, The Great War had degenerated into trench warfare and day-and-night artillery duels. That year the German Army employed poison gas for the first time. Chlorine, phosgene, and mustard gas caused slow agonizing death. Combatants lived and died in the squalor of trenches filled with knee-deep mud, urine, and feces, and they choked on stench from decaying corpses.

> Rats scurried about, feeding on the dead and on flesh of living men too weak to defend themselves.

A non-stop cannonade ripped apart human bodies, and shrieks of wounded men, writhing in filth and screaming for their mothers, drove some soldiers insane. Rats scurried about, feeding on the dead and on flesh of living men to weak too defend themselves. The rats carried lice, which in turn carried trench fever. Unspeakable horror, misery, exhaustion, and terror became the daily norm. Men standing face-to-face had to scream at each other to be heard over the unending cacophony of exploding artillery shells.

In this worldly equivalent of Dante's descent into Hell, LtCol. John McCrae (1872-1918), MD, Canadian Army, labored as a surgeon in the Canadian First Field Artillery Brigade. In his field hospital – a bunker filled with mud in the Ypres salient in Belgium – he treated wounded men from "Flanders' fields" as best he could. In a letter to his family he called the task "days of Hades."

> None can realize the horrors of war, save those actually engaged. The dead lying all about, unburied to the last. My God! My God! What a scourge is war!
> [Pvt. Samuel Johnson, CSA, in a letter, 1864]

On 2 May 1915 a German shell killed McCrae's former student and close friend, Lt. Alexis Helmer. The death had a profound effect on McCrae. The next day, despite the shelling, he crawled atop his bunker and watched the birds (which he called *larks*) flying through the cannonade. He looked down at scores of wild red poppies that had sprouted between rows of white crosses that marked hundreds of new graves, including the day-old grave of his friend. The poppies swayed and rocked to and fro in the gentle breeze that swept over the graves.

> He looked down at scores of wild red poppies that had sprouted between rows of white crosses that marked hundreds of new graves.

On a scrap of paper, McCrae scribbled a brief poem. He wrote not from his viewpoint. Instead, he wrote as the voice of soldiers who had been killed. He titled his short poem, *In Flanders Fields*. McCrae then shrugged, tossed the paper away, and crawled back into his medical bunker.

SgtMaj. Cyril Anderson, who had been crouching in a nearby trench, retrieved the crumpled-up paper and saved it. Several months later he mailed ***In Flanders Fields*** to newspapers in England. The poem was first published on 8 December 1915:

– In Flanders Fields –

In Flanders fields the poppies blow
Between the crosses, row on row,
That mark our place; and in the sky
The larks, still bravely singing, fly,
  Scarce heard amid the guns below.

We are The Dead. Short days ago
We lived, felt dawn, saw sunset glow,
Loved and were loved, and now we lie
  In Flanders fields.

Take up our quarrel with the foe.
To you, from failing hands we throw
The torch; be yours to hold it high.
If ye break faith with us who die
We shall not sleep, though poppies grow
  In Flanders fields.

Postscript: *In Flanders Fields* achieved worldwide acclaim as the most poignant and memorable plea ever written about warfare. Sadly, LtCol. McCrae never received recognition in person. Like hundreds of thousands of men from many nations, he failed to survive the war. He lost his life in France on 28 January 1918.

The American Warrior's Rules for Life

For each American Warrior, life and service constitute a worldly trust and responsibility. Yet, before one sets out to change the world he should start at home. The person he sees in the mirror each morning is the person with whom he should begin. The most crucial person with whom a warrior – *or anyone* – must live is himself, so he should conduct himself in a manner to ensure that he remains in good company. He will take a giant step in that direction if he heeds ***The American Warrior's Rules for Life***:

The most crucial person with whom a warrior must live is himself.

In matters of conscience, ignore the majority.

Dare to be different. If all think alike, none are really thinking.

Right or wrong, your *silence* equals your *consent.*

Avoid any philosophy prompted by a lack of courage.

Character is what you are in the dark.

Beware of the road to misery – trying to please everyone.

Character is what you are in the dark; your *character* is your *destiny.*

Right or wrong, your *silence* equals your *consent.*

When you always tell the truth, you never need to remember anything.

Believe in, and sacrifice for, *a cause greater than yourself.*

Reject any so-called reasoning which consists of searching for a rationale for believing what you already believe.

On 20 April 2007 an American Warrior befriends a young child in Fallujah, Iraq (photo courtesy of U.S. Marine Corps).

Remember that (1) small minds discuss *people*, (2) average minds discuss *events*, but (3) great minds discuss *ideas*.

Never argue with a fool – it's akin to wrestling with a pig.

Tilting at windmills hurts you more than it hurts the windmills.

Remember, no monuments are erected to honor cynics.

It's great to be great. It's *greater* to be human.

In interpersonal matters, apply The Golden Rule.

It's great to be great. It's *greater* to be human.

Never sneer at anyone's dreams, for dreams may be all they have.

Remember, the better part of your life consists of your friendships.

Never allow a little dispute to damage a great friendship.

Count your wealth by your friends, not your dollars.

Count your wealth by your friends, not your dollars.

Dare to dare, for nothing worthwhile is achieved without risk.

Judge an accomplishment by what you risked to achieve it.

The keys to success: (1) vision, (2) initiative, (3) commitment.

Each day, the best place to start from is where you are.

Yesterday is history. Tomorrow is never guaranteed. Today is all you have to work with.

Never give up. Never back down. Never give in. Keep scratching!

– The Challenge –

Life is always in session.
Are you always present?

Requiem:
Eulogy for American Warriors

The world's most famous photograph is Joe Rosenthal's image of the flag-raising atop Mt. Suribachi, Iwo Jima, on 23 February 1945. The most poignant ***Eulogy for American Warriors*** was delivered in the shadow of that volcanic massif over a month later. A brief historical review is warranted:

The Battle for Iwo Jima: During the closing years of World War II the Japanese knew their military base on Iwo Jima would be attacked. The tiny volcanic island was a mere flyspeck in the vast Pacific. Yet, it lay only two hours' flying time from Japan's home islands. Its two airfields were crucial to Imperial Japan.

On Iwo Jima, Japanese mining engineers used slave labor to dig and blast 16 miles of tunnels to link over 1,400 underground rooms and fortified fighting positions. Laborers built hospitals and supply rooms under hundreds of feet of solid rock. They dug and drilled huge caves into the igneous face of volcanic Mt. Suribachi. Inside the caves they mounted heavy artillery on roll-out tracks and protected these emplacements with reinforced steel doors. Japan had turned Iwo Jima into a seemingly impregnable fortress.

Japan had turned Iwo Jima into a seemingly impregnable fortress.

American bombing and shelling had little effect on the bastion. Nonetheless, on 19 February 1945, U.S. Marines stormed the beach. Iwo Jima became a charnel house, a bloodbath. The brutal struggle raged nonstop for 37 horrible days. The Marines finally prevailed, but they lost 6,821 men killed and 19,217 more maimed or grievously wounded.

Iwo Jima became a charnel house, a bloodbath. The brutal struggle raged nonstop for 37 horrible days.

The Eulogy: During the battle a Jewish rabbi from New York, chaplain Roland B. Gittlesohn, had served with the Marines. He was

In February 2005 a bugler sounds Taps atop the Marine Corps War Memorial to commemorate the 60th anniversary of the battle for Iwo Jima (photo courtesy of U.S. Marine Corps).

the first Jewish chaplain assigned to the Marine Corps. He had shared the fear and horror of combat, ministering to the wounded and praying for the dead. After the fighting ended the division chaplain selected him to deliver a eulogy during dedication of the Fifth Marine Division cemetery on the island.

Standing before thousands of fresh graves in the shadow of Mt. Suribachi and addressing the survivors, chaplain Gittlesohn spoke softly, yet eloquently. The *Associated Press* documented his somber ***Eulogy for American Warriors***. It has been entered, verbatim, into the U.S. Congressional Record. Gittlesohn's words, spoken long ago, remain a timeless tribute to each American Warrior who has sacrificed his life in an effort to further the cause of freedom:

– Eulogy for American Warriors –

Here before us lie the bodies of comrades and friends. Somewhere in this plot of ground there may lie the man who could have discovered a cure for cancer. Under one of these Christian Crosses, or beneath a Jewish Star of David, there may rest a man who was destined to be a great prophet. But now they lie silently in this sacred soil, and we gather to consecrate this earth to their memory.

Here lie men who loved America.

It is not easy to do so. Some of us have buried our closest friends here. We saw these men killed before our very eyes. Any one of us might have died in their places. Indeed, some of us are alive and breathing at this very moment only because the men who lie here beneath us had the courage and the strength to give their lives for ours.

Some of us have buried our closest friends here. We saw these men killed before our very eyes.

. . . Here lie men who loved America because their ancestors, generations ago, helped in her founding, and [here lie] other men who loved her with equal passion because they themselves, or their fathers, escaped from oppression to her blessed shores. Here lie officers and men, Negroes and Whites, rich and poor, together. Here no man prefers another because of his faith or despises him because of his color. Among these men there is no discrimination, no prejudice, no hatred. Theirs is the highest and purest democracy. To this, then, as our solemn and sacred duty, do we the living now dedicate ourselves – to the right of

Among these men there is no discrimination, no prejudice, no hatred.

Protestants, Catholics, and Jews, [the right] of white men and Negroes alike to enjoy the democracy for which all of them here have paid the price.

This war has been fought by the common man. Its fruits must be enjoyed by the common man. We promise, by all that is sacred and holy, that your sons, the sons of miners and millers, the sons of farmers and workers, will inherit from your death the right to a living that is decent and secure.

. . . Thus do we memorialize those who, having ceased living *with* us, now live *within* us. Thus do we consecrate ourselves, the living, to carry on the struggle they began. Too much blood has gone into this soil for it to lie barren. Too much pain and heartache have fertilized the earth on which we stand.

We here solemnly swear – this shall not be in vain!

We here solemnly swear – this shall not be in vain! Out of this, from the suffering and sorrow of those who mourn will come, we promise, the birth of a new freedom for the sons of men everywhere.

Amen.

– Words of Wisdom –

There is no room in this country for hyphenated Americanism! The one absolutely certain way of bringing this nation to ruin, of preventing all possibility of its continuing to be a nation at all, would be to permit it to become a tangle of squabbling nationalities.

[Theodore "Teddy" Roosevelt, former U.S. President, New York, 12 October 1915]

About the Author

Marion Sturkey began his military and aviation career in the U.S. Marine Corps. He earned the Naval Aviator designation in 1965 and flew both fixed-wing aircraft and helicopters. He became a Commercial Pilot after he left active duty, and for two years he flew helicopters to support off-shore oil exploration in the Gulf of Mexico.

Thereafter, Marion worked in various management capacities at BellSouth Corporation for 25 years. During the last ten years of that career he served as guest instructor at Bell Communications Research in Illinois and New Jersey. After retirement from the corporate world he began writing books about military and aviation concerns. He has been a frequent guest speaker at military bases and at functions for military veterans. He holds life memberships in the Marine Corps Combat Helicopter Association and the Marine Corps League.

Marion's first venture into military satire garnered acclaim for its timeless truisms and insight. This new *second edition* of that unique military classic, ***Murphy's Laws of Combat***, is his ninth book.

Books by the Author

BONNIE-SUE: A Marine Corps Helicopter Squadron in Vietnam
ISBN: 0-9650814-2-7 (509 pages)

Warrior Culture of the U.S. Marines (first edition)
ISBN: 0-9650814-5-1 (207 pages)

Warrior Culture of the U.S. Marines (second edition)
ISBN: 0-9650814-1-9 (212 pages)

Murphy's Laws of Combat (first edition)
ISBN: 0-9650814-4-3 (241 pages)

Murphy's Laws of Combat (second edition)
ISBN: 978-0-9650814-6-7 (366 pages)

MAYDAY: Accident Reports and Voice Transcripts from Airline Crash Investigations
ISBN: 978-0-9650814-3-6 (461 pages)

MID-AIR: Accident Reports and Voice Transcripts from Military and Airline Mid-Air Collisions
ISBN: 978-0-9650814-7-4 (477 pages)

GONE, BUT NOT FORGOTTEN: An Introduction
LCCCN 88-92649 regional historical interest only (42 pages)

GONE, BUT NOT FORGOTTEN
LCCCN 88-92560 regional historical interest only (679 pages)

(coming soon) *Military Monuments in South Carolina*
ISBN: 978-0-9650814-8-1

Index